Studies in Geography 1

Authors

Eddie Broadley

Linlithgow Academy, Linlithgow

Richard Goring

Brannock High School, Motherwell

General Editor

Ritchie Cunningham

Social Subjects Adviser, Highland Region

Oliver & Boyd

Acknowledgements

The publishers wish to thank the following for perrmision to reproduce photographs.

J. Allan Cash (A, B, p. 6; D, E, p. 7; B, p. 10; G, H, p. 13; C, p. 20; D, F, p. 21; C, p. 27; A, C, E, p. 34; A, p. 52; D, p. 62; F, p. 63; A, p. 70; E, p. 71; C, p. 75; B, p. 78; E, p. 83; A, p. 92); Barnaby's (C, p. 6; E, p. 21; B, p. 53; A, p. 64; C, p. 78; D, p. 109); Scottish Tourist Board (C, p. 10; C, p. 48; D, p. 88); R. Goring (A–E, p. 11; C, D, p. 31; D, p. 69; C, p. 104); Glasgow Herald and Evening Times (p. 17; A, B, p. 19; B, p. 48; D, F, p. 49; A, C, p. 50; F, p. 51); Glasgow District Council (C, D, p. 19; A, p. 48; D, E, p. 49); Photo Source (G, p. 21; B, p. 62; C, p. 72; A, B, p. 88; C, p. 91; A, p. 98); Liverpool City Engineers Department (A, p. 22; C, p. 23); W. E. Marsden (B, p. 22); Barrat Developments plc (C, D, p. 23); East Kilbride Development Corporation (B, p. 24); E. Broadley (C, p. 24; E, p. 25; A, p. 54; C, p. 62; A, B (1–3), p. 86; C, p. 87; B(1–3), p. 96); J. Topham (D, p. 27; B, p. 64; D, p. 71; B, p. 82; A, p. 90; A, p. 104); Aerofilms (B, p. 28; A, p. 29; B, p. 36; A, p. 46; D, p. 47; E, p. 61; D, p. 64; G, p. 77); Massey Ferguson (B, p. 30); Daks–Simpson (p. 33); J. Dewar (B, p. 34); Scotsman Publications (C, p. 37); John Cornwell (A, p. 38); National Coal Board (B, p. 38); W. E. Marsden (A, p. 40); Ray Green (C, p. 41); Timex (p. 42); Committee for Aerial Photography, University of Cambridge (F, p. 43); Guinness Brewing (C, p. 45); Scottish Opera (E, p. 49); Whitefriars Advertising (E, p. 55); Commission for the New Towns (B, p. 56); A. C. Waltham (E, p. 63; D, p. 76); Swiss National Tourist Office (G, p. 63; B, p. 80; E, p. 89; A, p. 94, C. p. 95); Camera Press (D, p. 73; C, p. 88); Airviews (M/C) Ltd Altrincham (C, p. 76); Bart Hofmeester, Rotterdam (B, p. 84); K. L. M. Aerocarto (D, p. 85); Alec Cargill (B(4), p. 96); Forestry Commission (p. 101); Ian Murphy (D, p. 103); South of Scotland Electricity Board (B, p. 104); North of Scotland Hydro-Electric Board (D, p. 104); Shell (A, p. 106);

The Ordnance Survey map extracts on pages 57 and 60 are printed with the permission of the Controller of Her Majesty's Stationery Office, Crown Copyright reserved. The map extract on page 51 is reproduced with the permission of Geographia Ltd, London. Crown Copyright reserved.

The authors wish to acknowledge the following sources of material, and individuals, for their help in preparing the text:

Scotts Porage Oats (Unit 7); Daks–Simpson (Unit 13); National Coal Board, British Petroleum (Unit 17); Clive Lewis (I.S.T.C.), Douglas Henderson (G.M.B.A.T.U.), Gavin Laird (A.U.E.W.), G. Stanley, British Aluminium (Unit 18); Cambridge Science Park (Unit 19); Guinness Brewing (Unit 20); London Docklands Development Corporation (Unit 21); Department of Employment Gazette, Falkirk Job Centre, Portal (Linwood) Ltd (Unit 25); Scottish Development Agency, Falkirk Job Centre (Unit 26); Blue Circle Industries Ltd (Unit 30); Hugh Johnson, *World Atlas of Wine* Mitchell Beazley 1971 (Unit 37); Thomson 'A la Carte' brochure, Eric Brown (Unit 42); Thomson 'Winter Ski' brochure, Isobel Brown (Unit 43); S.A.G.T. occasional paper 1985, Countryside Commission for Scotland (Unit 44); South of Scotland Electricity Board, North of Scotland Hydro-Electric Board (Unit 48); *The Economist* (Unit 50); Central Statistical Office in Lusaka, Vauxhall (Unit 52); Unilever, *The Economist*, Peter Donaldson *Economies of the Real World* Penguin 1981 pp. 227–30 (Unit 53).

The authors are also grateful to Anne Bainton, Morag Keenan and Nanette Menzies for remedial advice.

Illustrated by Julia Cobbold, John Marshall, John Lobban and Cauldron Design Studio.

Cover design by Cauldron Design Studio.

Oliver & Boyd
Robert Stevenson House
1–3 Baxter's Place
Leith Walk
Edinburgh EH1 3BB
A division of Longman Group UK Ltd

ISBN 0 05 004055 3

First published 1987

Set in 11/13pt Helvetica and Helvetica Bold

Produced by Longman Group (F.E.) Ltd.

Printed in Hong Kong

Contents

1 Where on Earth?

Every place on Earth is different.
Some places have not changed for thousands of years: for example, very high mountain areas, hot and cold deserts. These are called **natural landscapes**.

A City centre, New York

B An arable farm

C A town park

Most places, however, have been changed by people. Sometimes the changes are quite small, such as putting a fence on a hillside to stop sheep from straying.

Sometimes the changes are greater. People have divided the land into fields for farming, and planted crops on it.

Sometimes people have completely changed the landscape by building houses, offices, shops and factories on it. When we look around these places, most of what we see is made of brick, concrete, metal or glass. These places can be called **artificial landscapes** because almost none of the natural landscape is left.

1 Look carefully at photographs A–E. They show landscapes which range from artificial to natural. Draw a box and scale like the one below. On this scale, 1 means completely artificial, 2 means fairly artificial but with some natural features left, and so on. Decide how artificial or natural each of the landscapes in the photographs is. Write the letter (A–E) of each photograph under the scale number which you think it fits. One has been done for you.

Artificial ——————————— Natural				
1	2	3	4	5
				E

2 For each photograph, write a sentence to explain why you made the choice you did.

3 Extracts F–J are from some well-known books and poems. They describe different kinds of landscape. Read the extracts carefully and try to decide which number from the artificial–natural scale each would fit. In the scale box that you drew for question 1, write the letters F–J under the scale numbers which you think the extracts fit.

E Mountain peaks in the Andes

D A country village

F 'The valley of the Umzimkulu. The surrounding hills are green and well-tended: the valley and lower slopes are bare and over-worked.'

Alan Paton, *Cry, The Beloved Country*

G 'First a canal is cut across Brangwen Farm, and then a railway traverses it; and all the time the collieries approach, with their attendant redbrick house-rows.'

D.H. Lawrence, *The Rainbow*

H 'Salisbury . . . a sprawling, ugly, beautiful, modern city, with a history. There are suburbs, and slums, chimneys, and concrete towers; but the trees redeem.'

Doris Lessing, *Nine African Stories*

I 'The curfew tolls the knell of parting day,
The lowing herd winds slowly o'er the lea,
The ploughman homeward plods his weary way,
And leaves the world to darkness and to me.'

Thomas Gray, '*Elegy*'

J 'I heard a siren coming from the docks
Saw that train set the night on fire,
I smelt the spring on the smokey wind
Dirty old town; dirty old town.'

Ewan MacColl, '*Dirty old town*'

2 Here and there...

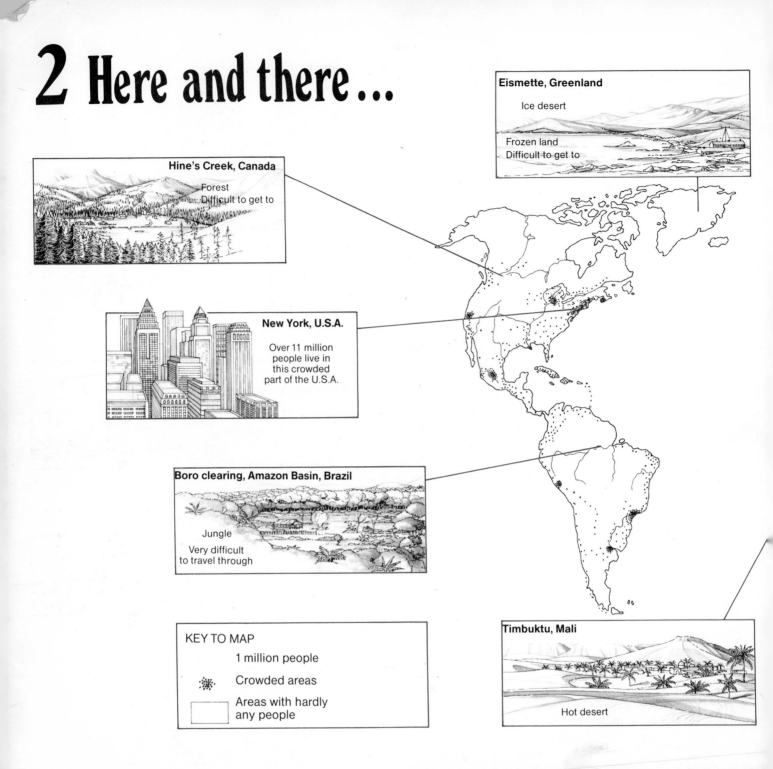

Eismette, Greenland

Ice desert

Frozen land
Difficult to get to

Hine's Creek, Canada

Forest
Difficult to get to

New York, U.S.A.

Over 11 million
people live in
this crowded
part of the U.S.A.

Boro clearing, Amazon Basin, Brazil

Jungle

Very difficult
to travel through

KEY TO MAP

1 million people

Crowded areas

Areas with hardly
any people

Timbuktu, Mali

Hot desert

Planet Earth is 'home' for almost 5000 million people. By the end of the century, there may be as many as 7000 million people.

Map A shows the areas of the world where most people live. This is called a **population distribution map**. Each dot on the map represents a million people.

As you can see, there are large areas where hardly anyone lives. These are **difficult areas** for people to live in: for example, hot and cold deserts, high mountains or huge thick forests.

On the other hand, there are some areas where many people live. Most of the world's population is found in **clusters**. Large clusters are usually in places that are (or were in the past) **easier areas** to live in.

What makes some areas easier to live in than others? There may be many reasons: for example, a pleasant climate, a fertile soil which is good for farming, flat land which is easy to build on, with water nearby. Other reasons might be to do with trade: for example, good transport routes, and important minerals such as coal, iron ore or oil.

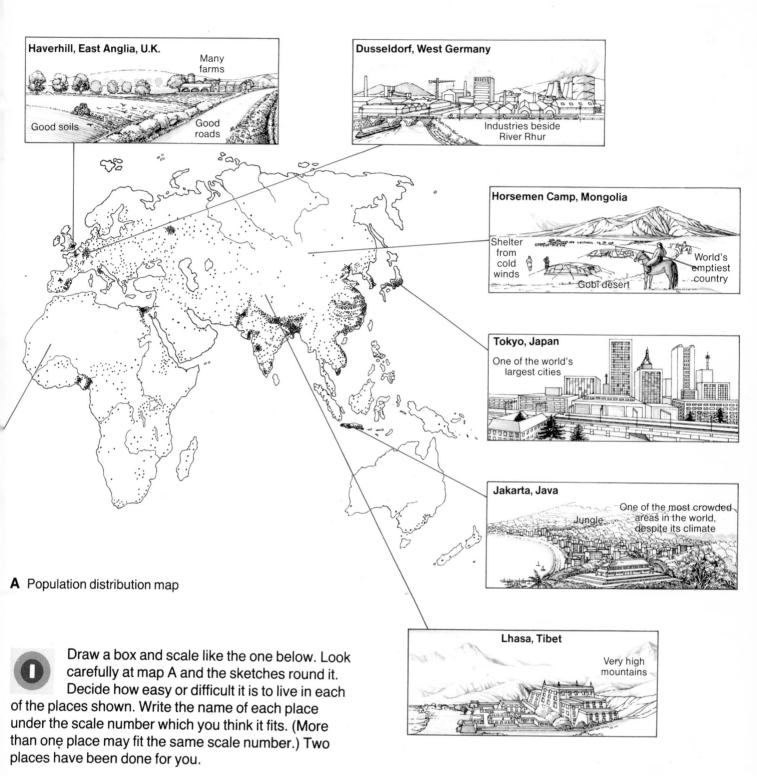

Haverhill, East Anglia, U.K.
Many farms
Good soils
Good roads

Dusseldorf, West Germany
Industries beside River Rhur

Horsemen Camp, Mongolia
Shelter from cold winds
Gobi desert
World's emptiest country

Tokyo, Japan
One of the world's largest cities

Jakarta, Java
Jungle
One of the most crowded areas in the world, despite its climate

Lhasa, Tibet
Very high mountains

A Population distribution map

1 Draw a box and scale like the one below. Look carefully at map A and the sketches round it. Decide how easy or difficult it is to live in each of the places shown. Write the name of each place under the scale number which you think it fits. (More than one place may fit the same scale number.) Two places have been done for you.

Easier to live in				Difficult to live in
1	**2**	**3**	**4**	**5**
New York				Lhasa

2 Copy out the following sentences, filling in the missing words and reasons by using the information in the sketches round map A, and the text.

The spread of people is not even all over the world. Some areas such as _____ are almost empty. This area is difficult to live in because _____. Other areas such as _____ have many people living in them. This area is easier to live in because _____.

9

3 Highs & Lows

Britain's population of 56 million people is not evenly spread over the whole country (see map A). Eight out of every ten people live in towns or cities (photo B). In these areas, on average, more than 200 people live on every square kilometre of land.

There are large areas of countryside and mountains where very few people live (see photo C). In these parts of Britain the **population density** is low: on average, there are fewer than six people living on every square kilometre of land.

1 Compare map A with an atlas map of Britain's highlands and lowlands. Where do most people live: in highland or lowland areas?

2 Photo C shows an area of low population density.
(a) What sorts of jobs might be found there?
(b) Why do you think few people live and work in this area?

B Croydon, London: an area where many people live and work (high population density)

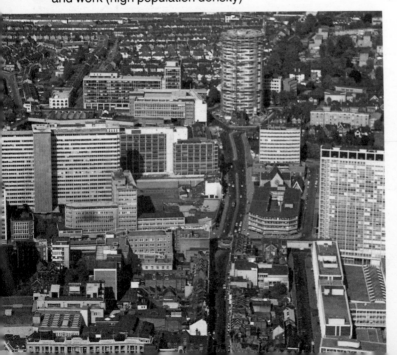

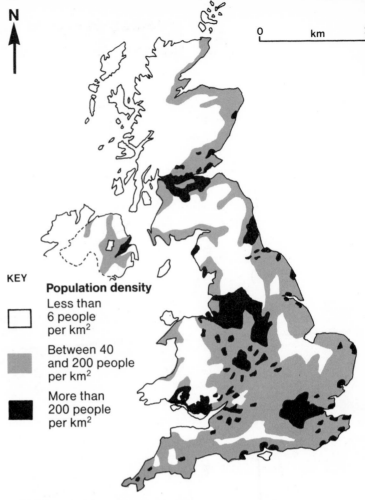

KEY
Population density

☐ Less than 6 people per km^2

▨ Between 40 and 200 people per km^2

■ More than 200 people per km^2

0 ___ km ___ 25

N ↑

A Britain's population density

3 Study photo B. Make a list of places that might provide jobs for people in Croydon (for example, offices).

4 Why do most British people live in towns and cities, not in the countryside? Give as many reasons as you can.

C An area where very few people live or work (low population density)

4 SETTLE DOWN!

George McNeill is production manager of Calchem Plastics in Bradford. The firm has recently had so many orders that it is going to open another factory, in Newcastle.

George has been made manager of this new factory. He, his wife Kate and son Paul will have to move to Newcastle when George starts his new job in a month's time. They are looking for a new house in the Newcastle area. An estate agent has sent them descriptions of five suitable houses, each in very different kinds of places (see adverts A–E).

The McNeill family has a difficult decision to make! The houses they are looking at are in different sizes of **settlement**. The smallest type of settlement is one house on its own and the largest type is a city. A settlement is a place where people live. Small settlements such as single buildings, hamlets and villages are found in the countryside. They are called **rural** settlements. Large settlements such as towns and cities are heavily built up with houses, shops, offices and factories. They are called **urban** settlements.

A

Near Ebchester, Co. Durham
A recently modernised, fully equipped former farmhouse in the countryside. In totally secluded surroundings, 8 km from Ebchester village and 40 minutes drive away from Newcastle. Situated in 15 hectares of woodland, with its own private road, this property boasts 4 bedrooms, 2 public rooms, solid fuel fires, double garage and several outhouses.

B

South Gosforth, Newcastle
A large semi-detached Victorian town-house in quiet residential suburbs, comprising three large bedrooms, lounge, large dining room kitchen, cellar and shared garden. Close to metro railway and all local services. Only 5 minutes from the city centre and urban motorway.

C

Belsay, Northumberland
This modern 5-apartment detached bungalow in a country village provides superb accommodation in a most attractive area. Close to the village shops and the church, the property enjoys magnificent views of the countryside. Three bedrooms, two public rooms, a large dining kitchen, garage and well-kept garden make this a most desirable property.

D

Durham, Durham Region
A traditional sandstone detached villa in a most sought-after part of this busy town. Close to all amenities including shops, railway station and schools. Five minutes from the motorway and 25 minutes from the centre of Newcastle, this four-bedroom, two-public room, fully modernised and equipped house is exceptionally good value. It includes a garage, mature garden and gas central heating.

E

Near Belsay, Northumberland
A fully up-graded, traditional detached house in a beautiful rural setting. One of a group of eight houses 4 km from the village of Belsay. Only 30 minutes drive from Newcastle, this property combines rural life with easy access to the city. The house has three large bedrooms, lounge, dining room, fully modernised kitchen, gas central heating, a large garage, well-maintained gardens and ½ hectare rough land.

The McNeill family will have to look at the advantages and disadvantages of living in each of the different settlements, as chart F shows. One of the things that may be important to them is what **services** (for example, shops, schools, public transport, sport centres) the settlements have. Large urban settlements can provide a wide variety of services. Small rural settlements may have very few.

F Type and size of settlement

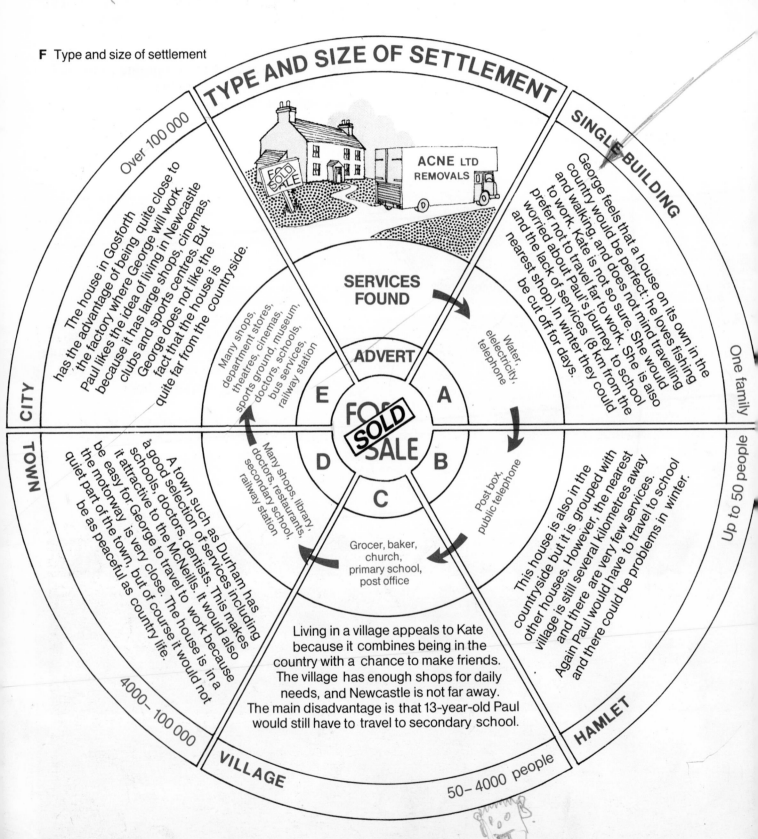

TYPE AND SIZE OF SETTLEMENT

SERVICES FOUND

ADVERT

SOLD
FOR SALE

E   A
D   B
C

Over 100 000

The house in Gosforth has the advantage of being quite close to the factory where George will work. Paul likes the idea of living in Newcastle because it has large shops, cinemas, clubs and sports centres. But George does not like the fact that the house is quite far from the countryside.

CITY

TOWN

A town such as Durham has a good selection of services including schools, doctors, dentists. This makes it attractive to the McNeills. It would also be easy for George to travel to work because the motorway is very close. The house is in a quiet part of the town, but of course it would not be as peaceful as country life.

4000 - 100 000

VILLAGE

Living in a village appeals to Kate because it combines being in the country with a chance to make friends. The village has enough shops for daily needs, and Newcastle is not far away. The main disadvantage is that 13-year-old Paul would still have to travel to secondary school.

50 - 4000 people

Many shops, department stores, theatres, cinemas, sports ground, museum, doctors, schools, bus services, railway station

Many shops, library, doctors, restaurants, secondary school, railway station

Grocer, baker, church, primary school, post office

Water, electricity, telephone

Post box, public telephone

SINGLE BUILDING

George feels that a house on its own in the country would be perfect: he loves fishing and walking, and does not mind travelling to work. Kate is not so sure. She would prefer not to travel far to work. She is also worried about Paul's journey to school and the lack of services (8 km from the nearest shop). In winter they could be cut off for days.

One family

Up to 50 people

This house is also in the countryside but it is grouped with other houses. However, the nearest village is still several kilometres away and there are very few services. Again Paul would have to travel to school and there could be problems in winter.

HAMLET

1 Read adverts A–E on page 11 carefully. Decide what size each settlement is. Write the letters A–E in order from the smallest to the largest settlement.

2 (a) which of the adverts A–E describe houses in rural settlements?
(b) Which of the adverts decribe houses in urban settlements?

3 Look carefully at photograph G. Make a list of all the things you can see which tell you that this shows an urban area.

4 Make a list of all the things you can see in photograph H which tell you that this shows a rural area.

5 Look carefully at chart F on page 12. Make a larger copy of the following table and complete it by listing the services that each of the settlements has.

Settlement	Services
Single building	
Hamlet	
Village	
Town	
City	

6 Look at chart F again. If you were a member of the McNeill family, which of the houses would you choose as your new home? Write a few sentences to explain your choice.

7 Write a few sentences to explain why large settlements have more services than small ones.

G An urban landscape

H A rural landscape

5 BUILD IT HERE !

A

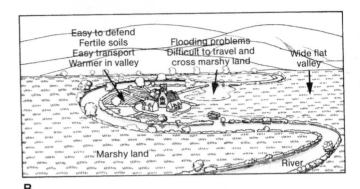

B

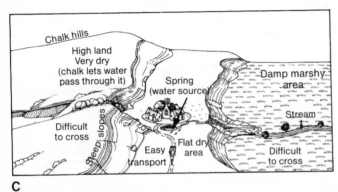

C

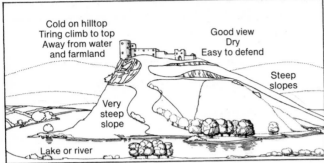

D

Why did settlements grow up in certain places? Usually it was because these places were suitable to build on. The exact place in which a settlement is built is called its **site**. Some sites are easy to build on and live on: for example, flat land with water nearby. Others can be difficult: for example, mountainous land or marshy land. Look carefully at pictures A–D. They show some different types of site.

All settlements in Britain started when people chose a site to live on. In the past the site had to be chosen very carefully if both the settlement and its people were to survive. The safer, drier and easier the land was to build on, the better chance that settlement had of growing.

The name of a settlement may give a clue to its age and why the site was chosen. The best clue is usually found at the end of the place name.

Roman times Words ending in	Meaning
-caster, -cester -port	castle harbour or gate
Viking times Words ending in	Meaning
-beck -dale -ings -by -kirk -toft, -thwaite	stream valley marsh homestead church clearing

Anglo-Saxon times Words ending in	Meaning
-ing -ham -ton -ley, -leigh, -field -mere -ey, -ea -hurst, -hirst, -holt, -wold	land belonging to a family homestead farm clearing lake island wood
Gaelic times Word ending (or beginning) in	Meaning
-dun -coille -cul -allt, -strath	fort wood shelter stream, valley

① Look carefully at pictures A–D. For each picture, say whether each site is good or bad, and give your reasons.

② Sketch E shows eleven settlements (labelled a–k). The site for each settlement was chosen for a particular reason. Copy the following table into your notebook and fill in each of the settlement letters a–k opposite the reason why it was built on that site. One has been done for you.

Settlement	Reason why site was chosen
h	Coal mining area In forest clearing Easy to defend On an island Beside a lake Good coastal fishing area Dry mound in marshy area Good water power Beside a spring Good flat land with water nearby At furthest point upriver for ships

③ Sketch E also shows six possible sites (labelled 1–6) for other settlements. In your notebook write the numbers 1–6 in a list. Beside each one write the description which best explains whether that site would be good to build on or not. One has been done for you on the sketch. For the other five, choose from the following.

Easy to defend this site   Flooding likely here

Too high, cold and wet   Slopes too steep to farm

Good fertile farmland

④ (a) Using the tables of place name endings on page 14 to help you, decide which of the settlements a–j is most likely to have each of the names shown below. Write the letters and names in your notebook.
Dunhill, Longmere, Broadley, Lowbeck
(b) Again, using the table of place name endings to help you, make up names for the following settlements on sketch E: a, b, e, f.

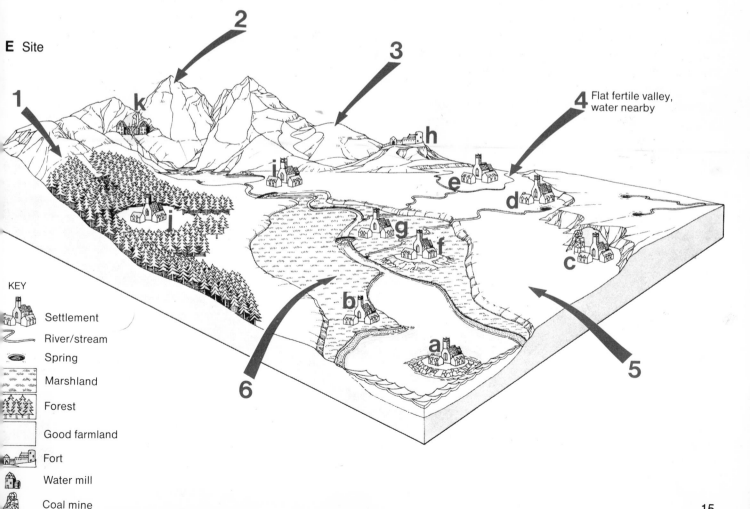

E Site

KEY

🏘 Settlement

〰 River/stream

◉ Spring

Marshland

🌲 Forest

Good farmland

🏰 Fort

🏭 Water mill

⛏ Coal mine

15

6 Whereabouts?

Look carefully at sketch E on page 15 and sketch A below, which shows the same area some years later. What changes have taken place? Why do some settlements grow bigger and more important than others? The answer has to do with where the settlements are, and how easy they are to get to. No matter how good the site of a settlement may be, the settlement is unlikely to grow into a large town or city unless it has a good **situation** (whereabouts) as well.

A

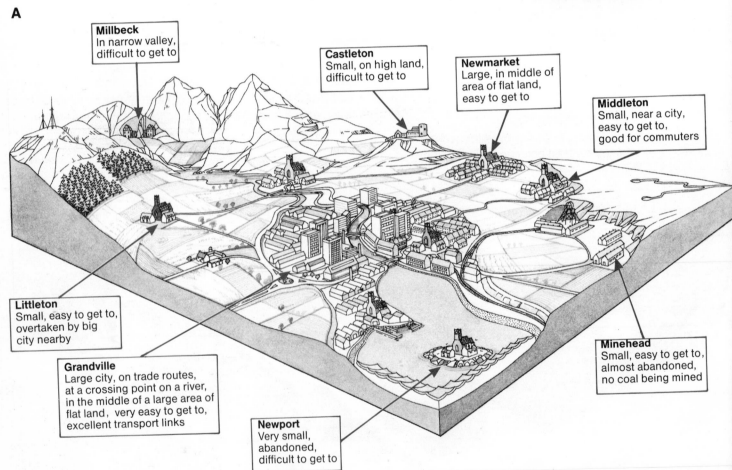

Millbeck
In narrow valley, difficult to get to

Castleton
Small, on high land, difficult to get to

Newmarket
Large, in middle of area of flat land, easy to get to

Middleton
Small, near a city, easy to get to, good for commuters

Littleton
Small, easy to get to, overtaken by big city nearby

Grandville
Large city, on trade routes, at a crossing point on a river, in the middle of a large area of flat land, very easy to get to, excellent transport links

Newport
Very small, abandoned, difficult to get to

Minehead
Small, easy to get to, almost abandoned, no coal being mined

As **trade** and **industry** developed, some settlements (such as Grandville) grew into large important cities because goods could easily be brought in and out by rail, road or sea. These **trade routes** gave the settlement a good situation. Other settlements did not grow because they were difficult to get to. The easier a settlement was to get to, the more important it became. London's situation must have been excellent!

1 Look at sketch A. In your own words explain why some settlements grow larger while others stay small or are abandoned.

2 Describe the situation of the settlement where you live.

7 City News

If it's in the news – it's in the *News!*

Parking fees set to rise

Strathclyde Regional Council are set to increase parking fees in Glasgow by 25 per cent in April. The increase will affect car parks and parking meters. An official commented "Maybe this will encourage people to use public transport and reduce city-centre traffic-jams."

Urban motorway pile-up – 5 hurt

The thick fog which blanketed the city today was responsible for several road accidents. Worst was a 10 car pile-up on the M8 at the Kingston Bridge, in which 5 people were injured.

Full report – page 6

New plan for city shopping centre

by Kevin Ferguson

A major new plan for Glasgow's busiest shopping area was announced today by the city's planning department. The development, which will cost over £15 million, is designed to make the east end of Argyle Street an up-market shopping area, with a covered retail area, air-conditioning, fast-food shops, rest areas, coloured fountains and a wide range of shops. Traders in the area had mixed reactions. Several felt that the increase in shopping space would attract more people to the area, others were concerned at the probable increase in rent and rates. A council spokesperson said "It is our aim to make Glasgow a more attractive place for all shoppers – whether they are local people, office workers during their lunch breaks, or people from outside the city."

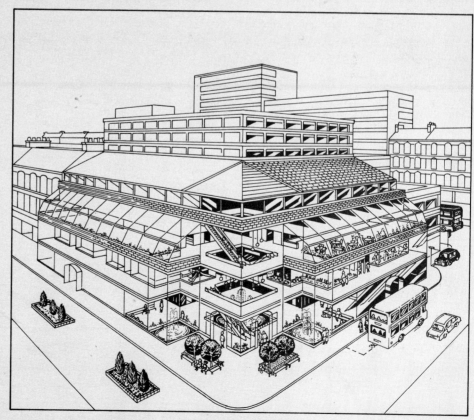

▲An artist's impression of the new Argyle Centre

◄Lunchtime shoppers in Argyle Street today

Residents complain – factory smoke gets in our eyes . . .

Householders in the Govan area of the city have complained to Health Authorities about thick smoke from Nelson Chemical Works. Spokesperson Mrs Angela Allan told our reporter: "There used to be dozens of factories in this area, but now Nelson's is all that's left. We need the jobs but not the pollution."

Health Authority experts have promised to investigate the complaint.

Council says No!

Permission to build a new private housing development in the Cathkin area of the city has been refused to Sherlock Homes Ltd, as it would extend into farmland which is part of Glasgow's Green Belt.

People's Marathon –
route finally agreed

Glasgow City Council today released details of this year's People's Marathon, to take place on 4 June. The route will take competitors through a wide variety of Glasgow's scenery, as shown in the map.

The first 21 km of the 42 km course will cover the north side of the River Clyde, before crossing George V Bridge and completing the distance on the south side of the River.

Last year's winner, Colin Cook, was enthusiastic about the route. "It's always nicer to run through different types of scenery", he said today, "It takes your mind off the pain."

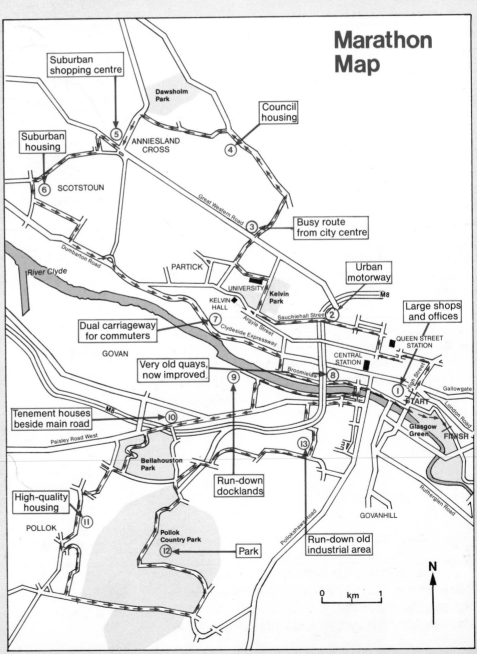

The articles in the *City News* give a clue to some of the ways the land in a city is used. Several different types of **land use** are found in and around a city.

Shopping land use includes all shops, from small corner-shops to large department stores.

Office land use is found mainly in the city centre.

Residential land use means all housing, from very small flats to large detached houses.

Industrial land use means land used for factories, warehouses, workshops.

Derelict land/waste land is unused and can include factories and houses which are now empty.

Transport land use includes land used by roads, railways and sometimes rivers or canals.

Recreational land use means land used for enjoying your free time or for sports activities: for example, parks, football grounds or golf courses.

Green belt is the countryside found round the edge of a city, which cannot be built on.

1 Copy the table on the right. Land use is often mentioned in the articles in the *City News*. Read each article carefully. Each time one of the land uses in the table is mentioned, put a tick in the middle column of your table. (Include the 13 places numbered on the map of the Marathon route.) When you have done this for all the articles, add up the ticks in each box to give a total for each land use.

Type of land use	Number of mentions	Total
Shopping		
Office		
Residential		
Industrial		
Derelict/waste land		
Transport		
Recreational		
Green belt		

2 The map of the Marathon route shows how the runners will pass through a wide variety of landscapes. Copy the table on the right. Each of the photographs on this page shows one of the numbered places on the map. Look very carefully at the map and each photograph, and try to complete the table.

Photograph	Site number
A	
B	
C	
D	

8 Around the City

In many large towns and cities, different **land uses** are found grouped together in the same area. Diagram A shows how land uses are grouped in a typical British city.

Large shops and offices, for example, are usually found in the **city centre** (1). The rest of the city has grown up around the centre.

During the 1800s, many factories and workers' houses were built around the city centre. Nowadays, many of these buildings have been demolished, are derelict, or are being improved. This area is often one of **housing renewal and industrial change** (2).

Large areas of **housing** of all kinds are usually found farther away from the city centre (3).

The government can stop new building round the edge of the city by making an area of **green belt** (4). This is done to stop cities growing too large and covering huge amounts of countryside.

Many cities have **lines of transport** (5): the main roads, railways lines or waterways where industrial land use is common. Factories need to be able to transport goods in and out of cities.

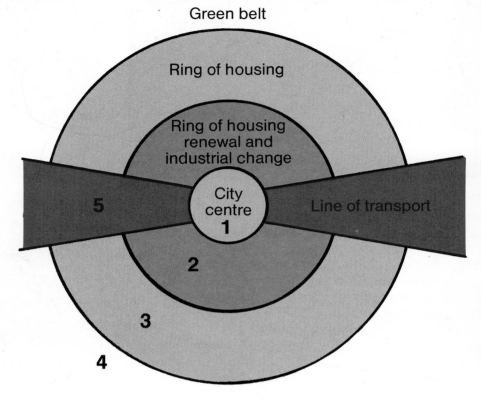

A Land use in a typical British city

Green belt

Ring of housing

Ring of housing renewal and industrial change

City centre 1

5   Line of transport

2

3

4

C

B Cross-section of a typical British city

a   b   c

D

E

F

G

① Look carefully at the Ordnance Survey map of Leeds on page 57. Copy the table below and complete it by matching each of the types of land use with a grid square on the map. Choose from: 2740, 3033, 2731, 3137, 2733.

Land use	Grid reference
City centre	
Housing renewal, industrial change and transport	
Housing	
Line of transport	
Green belt	

② Look carefully at the cross-section of a city (diagram B). Match each of the land uses 1–5 in diagram A to the correct lettered areas in the cross-section.

③ Look at photographs C–G. Match each of the photographs to the correct numbered part of diagram A.

④ Look at the photographs of Glasgow on page 19. Match each of the photographs to the correct numbered part of diagram A.
If you find this easy, try to do the same for each of the numbered places on the Marathon route map on page 18.

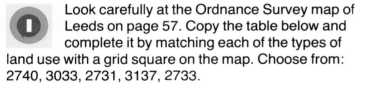

e   f   g   h

9 WEAR and TEAR

The old run-down areas of towns and cities are usually close to the city centre. The buildings in these areas were built during and after the **Industrial Revolution** in the early and mid 1800s. At that time towns and cities began to grow quickly, to house workers for the new factories.

Now many of these buildings have been demolished and those that haven't are near the end of their useful life (see photo A). There are signs of wear and tear to be seen in and around the old houses and factories (see photo B). This is called **urban decay**.

A

B

There are many reasons why these areas become run-down:

- very old or poorly designed buildings;

- houses with outside toilets, no bathrooms, no hot water and no gardens;

- overcrowding, with large families often living in one or two rooms only;

- old factories closing down and many people losing their jobs;

- lack of money for house repairs and improvements.

1 Copy the table on the right. Look carefully for signs of wear and tear in photos A and B. Complete your table by putting a tick in the correct box for each sign that you can see.

2 Add another two signs of wear and tear to your table. Which photo shows more signs of wear and tear?

Signs of wear and tear	Photo A	Photo B
Broken windows		
Boarded or bricked-up windows		
Boarded or bricked-up doors		
Damaged roof		
Part of building not lived in		
Broken gutters and drainpipes		
Waste land around building		
Totals		

Today some of the old run-down areas are showing signs of **recovery**. The problems of wear and tear are being tackled in different ways.

• **Tearing down and starting again.** Slum buildings can be demolished and new buildings built in their place. In many big cities, the tearing down of old houses and empty factories began over 40 years ago. In some places, high blocks of flats were built to replace the old houses. In others, the people in the run-down areas were moved to different areas altogether. The government gave money to help councils to build large new housing estates near the edge of cities for these people to live in. Some completely new towns were also built outside the old cities.

• **Improvement.** Many old areas can be improved without knocking down all the buildings (see photo D). Improvements to houses can mean a new roof, a new inside toilet, new paintwork and rewiring. Buildings can be cleaned up and new gardens made.

C

D

Signs of recovery	Photo C	Photo D
New windows		
Repaired roof		
Repaired stonework or walls		
New fence and walls		
Trees and grass planted		
Parking areas built		
New gutters		
Totals		

3 Copy the table on the right. Look carefully for signs of recovery in photos C and D. Complete your table by putting a tick in the correct box for every sign that you can see.

4 Add another two signs of recovery to your table. Which photo shows more signs of recovery?

10 New Living Room

One answer to the problems of wear and tear in city centres has been to build completely **new towns**. Today there are 28 new towns in Britain (see map A). East Kilbride in Scotland was one of Britain's first five new towns (see photo B). Work started on building it in 1947.

Living in a new town is very different from living in an old run-down city centre. New towns are carefully planned. Modern housing areas are built in attractive surroundings, with open play and sports areas nearby. Work can be found in the industrial estates on the outskirts of the new towns. These are built away from housing, close to good transport links. New town centres usually have tall office blocks, large car parks and covered shopping areas.

Although new towns are pleasant places to live and work in, there are some problems. Often there are few entertainments such as discos, cinemas or pubs. The long distances between houses, work and shops mean that people need a car to get about easily.

A Britain's new towns

B East Kilbride

C New Town advert

❶ Look at photo B. Write down the ways in which a new town is different from an old town.

Large **council housing estates** were built on the outskirts of big cities as another answer to the housing problems of the old inner-city areas. For example, Muirhouse on the north-west edge of Edinburgh (see map D) and Easterhouse on the north-east edge of Glasgow were built over 30 years ago. Thousands of people from the worst **slum areas** in the cities were moved to these and other housing estates.

The type of housing on these estates is different from the run-down city centre areas (see photo E). There are **high-rise flats**, **low-rise flats** and **terraced houses** with gardens. These modern houses have indoor toilets, bathrooms, and hot water. The estates also have open spaces, good roads, local churches and schools.

Some estates have been more successful than others. Unfortunately, many had to wait a long time for local services such as shops and libraries. People complained that they were too far away from shops and

entertainments of the city centre, and jobs were hard to find locally. In some estates, the buildings began to suffer from problems caused by dampness. Today many of these estates have become problem areas.

D

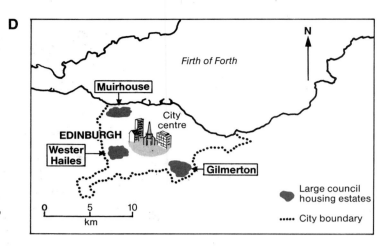

E Muirhouse, Edinburgh

F Sketch of photo E

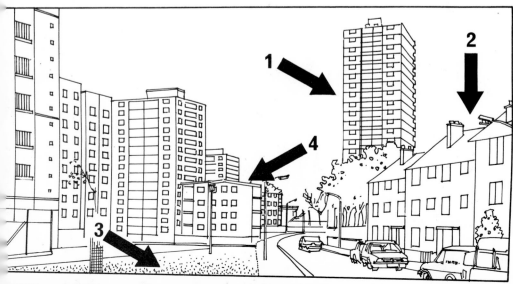

G

2 Copy the outline of sketch F, its title and the numbered arrows. Read this page again carefully. Add four labels to your sketch to describe what arrows 1–4 point to.

3 What do the newspaper headlines in picture G tell you about council estates like Easterhouse?

11 THE UGLY CITY

Many people like to live in cities, for different reasons. Cities are the centres of big business, there are lots of shops, and entertainments such as cinemas, clubs, sports centres. Also, in a city there are always plenty of other people around.

There are many attractive things about city life but, as the pictures on these pages show, there are many ugly things too.

Problems such as traffic jams, theft, vandalism, drug addiction, pollution, riots and shop-lifting are found in lots of places, but cities seem to get more than their fair share. Why is this? It is partly because cities have a great many people living in them. It may also be because the types of **environments** (or surroundings) in cities make certain problems more likely to happen.

The city environments of busy streets, alleyways, subways, run-down buildings, high-rise flats, slum areas and wasteland are easy places for trouble to break out.

B

Road accidents **Traffic jams** **Pollution** **Noise**	**1** Problems like these are found in all cities because they are busy places with lots of traffic and pedestrians. Using public transport can help reduce some of the problems, but this is not always a popular solution.
Racial tension **Vandalism** **Gang violence** **Muggings** **Football hooliganism** **Crime**	**2** All these problems are to do with breaking the law. Some are the result of organised crime, others happen because some people living in cities do not feel satisfied with their lives. They may be bored, poor, and angry against the society in which they live.
Drug dependence **Suicide** **Loneliness** **Down-and-outs** **Beggars**	**3** There are many reasons why these problems are particularly serious in cities. Large populations, poverty, poor housing, high unemployment, and unhappiness are just some of the reasons.

1 Write the heading 'Some problems in our cities'. Look carefully at these two pages and make a list of as many city problems as you can. Use the pictures and text to help you.

2 Look at arrow 1 in diagram B. How could more use of public transport help reduce some of the problems mentioned? What other things could be done to reduce the problems?

3 Look at arrow 2 in diagram B. Write a paragraph to describe some of the ways these problems might be helped.

4 Look at arrow 3 in diagram B. Choose **two** of the problems mentioned and write down some ways each might be helped.

5 Why do you think many people in cities are lonely, even although there are always plenty of other people around?

E

High-rise flats which are 20 years old are to be demolished because the cost of repair is too much

EVENING POST

Youths hold up bus driver at knife point
Passengers terrified
£40 stolen

Sir - It is surely about time the City Council did something about the terrible lack of parking facilities for cars in the city centre. On Thursday it took me almost an hour to find a parking space.

Sir - I wish to complain about the amount of litter which is messing up the centre of our city. Is it not about time that

C

Problems in our cities

D

S GO HOME
UNITED   GIZZAJOB
KINGS
RULE
O.K.

Central Parish Church

Samaritan Centre

lodgings for the homeless

12 Downtown?

Look at the Ordnance Survey map extract of Leeds on page 57. Part of this area is shown on photograph A opposite.

Like all British cities, Leeds has changed dramatically over the last 30 years. At one time, a typical scene was rows and rows of back-to-back terraced houses (photo B), like Coronation Street on television. Many of these have now been knocked down. They have been replaced by modern flats, low-rise houses, new industrial estates or sometimes just left as areas of **derelict land**. Where old houses remain, they are being improved.

1 List four features shown on the O.S. map which tell you that Leeds city centre is in grid squares 2933 and 3033.

2 What features on the map suggest that Leeds is an industrial city?

3 The housing shown in photo B is found in grid square 2733 on the map. Describe the housing, mentioning type of house, building material, fuel used, open space and street pattern.

4 The houses shown in photo B were built in the late 1800s. So were those in grid square 3238. What differences would you expect to find?

5 Describe the pattern of roads in the map extract. (Hint: where do they lead; are there any ring roads?)

6 In which grid square would you find
(a) the main railway station?
(b) Leeds University?
(c) a reservoir?
(d) Headingley cricket ground?

To answer questions 8–11, look carefully at the aerial photograph on the next page (A) and the O.S. map extract.

7 (a) In which direction is the photographer facing?
(b) What features in the photograph help you to recognise the city centre of Leeds?

8 (a) What features in the photograph do you think have not changed for at least 30 years?
(b) What features in the photograph would not have been there 30 years ago?

9 In which grid square would you find each of the features numbered 1–5 on the aerial photograph?

10 Write a paragraph describing how well the model of 'land use in a typical British city' (see page 20) fits Leeds.

B Traditional terraced houses in Leeds 1966 (Wortley)

A The centre of Leeds

13 What's in a job?

At some time in their lives, most people will work in a job. There are millions of different types of job, but all of them are linked together in some way. We can call these linked jobs 'job chains'.

Here is an example:

A

First: grain is grown on a farm.
↓
Second: the grain is milled to make flour and the flour is made into bread.
↓
Third: the bread is sold in shops.

All jobs fall into one of three main types of industry.

1. Jobs which involve harvesting crops from the land or sea, or extracting something from the earth are all **primary industries**. 'Primary' means 'first'. Jobs in primary industries are always first in the job chain.

2. The products of primary industries, such as grain, fish or oil, are often made into new products. For example, grain is ground to make flour, fish may be dried to make fishmeal, and oil may be refined to make petrol and other chemicals. Industries which make or manufacture products are all **manufacturing industries**. Jobs in the manufacturing industries are often second in a job chain.

B Primary industry

3. The third type of industry is **service industry**. This includes jobs which involve giving a service to people, such as nursing, hairdressing or teaching. It also includes jobs to do with the products of primary and manufacturing industries: for example, selling, repairing or advertising goods.

Here is a flow diagram to show how the three types of industry are linked together.

E | Primary | → | Manufacturing | → | Service |

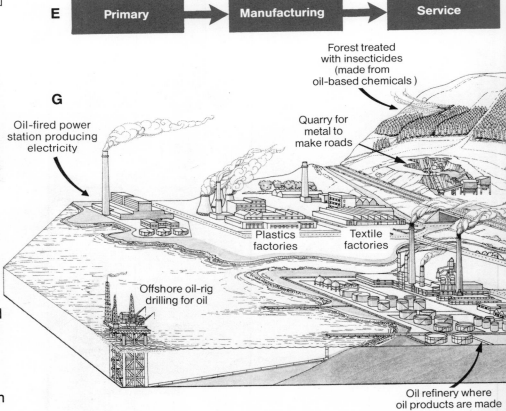

G

Oil-fired power station producing electricity

Forest treated with insecticides (made from oil-based chemicals)

Quarry for metal to make roads

Plastics factories

Textile factories

Offshore oil-rig drilling for oil

Oil refinery where oil products are made

C Manufacturing industry

D Service industry

The links are usually more complicated than this, because several different products and jobs may be involved. Here is another example:

F

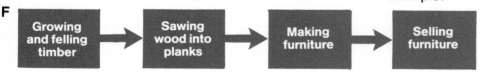

Growing and felling timber → Sawing wood into planks → Making furniture → Selling furniture

1 Here is a list of four jobs. Look at diagram A and then draw a job chain to show how the four jobs link together. Put each job in the correct order in the chain, and use arrows to make the links.

Baker, farmer, shop assistant, miller.

2 Copy and complete the table below by putting each job into the correct column.
Farmer, shop assistant, banker, shepherd, doctor, steelworker, fisherman, policeofficer, carpenter, tailor, forester, jeweller.

Primary	Manufacturing	Service

3 Choose a product, then draw a flow diagram similar to the one in F to show the main stages involved in producing and selling your product. Use different colours for your boxes to show which type of industry each stage belongs to.

4 Look at diagram G and read the labels. There are links between many of the things that you can see in diagram G. Try to make two or three flow diagrams, linking the labels together in a chain.

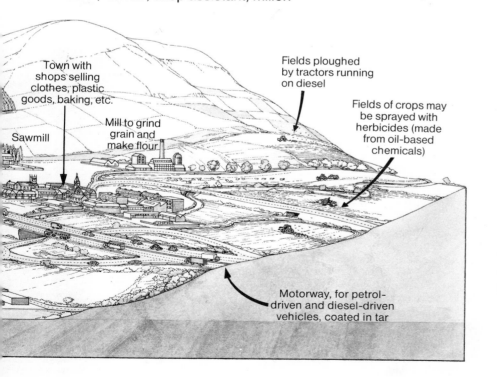

Town with shops selling clothes, plastic goods, baking, etc.

Fields ploughed by tractors running on diesel

Fields of crops may be sprayed with herbicides (made from oil-based chemicals)

Mill to grind grain and make flour

Sawmill

Motorway, for petrol-driven and diesel-driven vehicles, coated in tar

5 Read each of the following descriptions of jobs taken from a newspaper. Decide whether each describes a job in a primary, manufacturing or service industry. Make a larger copy of the table at the bottom of the page, and use it to record your answers.

SITUATIONS VACANT

1 '... will be expected to care for a dairy-herd in excess of 250 head. Accommodation is provided with the job.'

2 'previous experience in vehicle assembly is required, and the successful applicant will need to work as part of a team on the final stages of the production line.'

3 '...self-motivated presentable person at least 25 years of age, who will provide shorthand, typing and general services to the management. Previous office experience essential.'

4 '...persons willing to train in tree-work. Applicants must be physically fit and willing to work at heights.'

5 '... as one of a team of 10 producing a wide range of wholemeal products for local shops. No qualifications are required, but the successful candidate must be prepared to train on the job.'

6 '... no qualifications necessary. We are looking for someone who enjoys meeting people, and who will work hard to help us develop our business with the retail trade. Must own a car, be prepared to work without supervision, on a commission basis.'

7 '... experience in anoraks and jackets essential. Excellent piece-rates paid – the more you make, the more you make!'

8 'The successful applicant will work hard, two weeks off-shore at a time, in all weather conditions. Rig facilities include excellent food and recreation areas, plus own room. Top wages paid.'

9 'The work will involve making deliveries to our depots in the West Midlands. Applicants must hold a Heavy Goods Vehicle (Class 1) licence. Early morning starts are required.'

Job no.	Type of industry	Guess the job
1		
2		
3		
4		
5		
6		
7		
8		
9		

14 Making a living

When we think of an **industry** we usually think of a factory making something which will then be sold for profit. To do this, the factory owner has to buy **raw materials** which are made (manufactured) into a **finished product** by the factory **workforce**. This is shown in the simple diagram below.

Input → Process → Output

Raw materials → Factory → Finished product

In real life, things are not as straight-forward as that. The example of Daks–Simpson, a clothing firm, gives us a good idea of how a **manufacturing industry** works.

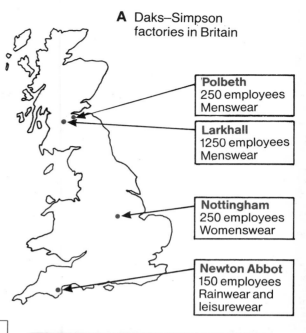

A Daks–Simpson factories in Britain

Polbeth
250 employees
Menswear

Larkhall
1250 employees
Menswear

Nottingham
250 employees
Womenswear

Newton Abbot
150 employees
Rainwear and leisurewear

Daks facts!

- Daks clothes are made under licence in Japan, South Korea, Canada and the U.S.A.
- The Larkhall factory alone makes about 8000 jackets and 15 000 pairs of trousers a week.
- Daks kit out Wimbledon umpires and Glasgow Rangers football club.

DAKS Simpson
01-734 2002 PICCADILLY

Workforce
Workers are trained by the company to become highly skilled

The Factory
Here, the cloth is cut into shapes before being sewn into garments. Many garments are hand-finished, then inspected. This means that the product is exclusive and therefore expensive.

Raw materials
Mostly British wools, especially Harris tweed. Cashmere is also used. Cotton thread, buttons and suede are also raw materials.

Pure new wool

Power, in the form of electricity, is used to heat and light the factory and drive the machines.

Finished clothes are transported directly by road to

DAKS-Simpson's only shop in Piccadilly, London

Major British stores such as House of Fraser, Rockhams, Dingles and Jenners

Marks and Spencer, for whom Daks make clothes

European and Asian countries. Over a third of the output goes to countries like France and Italy.

1 Look at map A. How many employees are there in Daks–Simpson factories in (a) Scotland, (b) England, (c) Wales, (d) Britain as a whole?

2 (a) List the inputs of a Daks factory.
(b) Make a list of the inputs, processes and outputs of a chocolate factory.

3 Find out what the following are: tweed, Harris tweed, cashmere, suede.

33

15 Pick a Spot

Most industries try to make as much profit as possible. This means that the owners must choose the place (or **location**) of the factory very carefully. The things which affect the owner's decision about the location of the factory are called **location factors**. Photographs A–E show some of the main location factors which are important in deciding where factories are built.

A The **government** will often help industries (in some areas) with special grants of money

- With Development Area, Enterprise Zone and Steel Opportunity Area status, Corby is able to offer a range of incentives that are second to none
- Corby has E.E.C. aid available for small businesses
- In Corby there is professional advice from day one to ensure that every new company obtains the best financial package that is available

B A **market** (a large town, for example). where the output products can be sold, should be near the factory. A town also supplies a workforce which is important to most factories

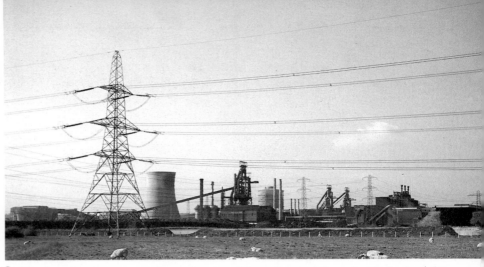

C **Fuel and power**: some factories need lots of electricity or bulky fuel such as coal

D **Site:** many factories need to be built on flat land, sometimes near water

E **Transport** is important for both inputs and outputs (finished products). Raw materials which are heavy or bulky should be close at hand

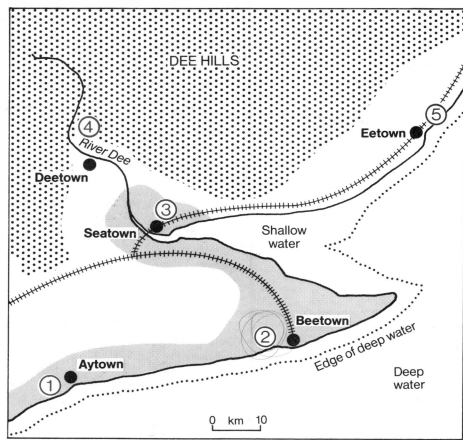

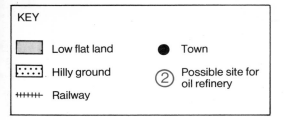

Black Oil company have decided to build a new oil refinery in the area shown in map F. An oil refinery is a very large factory which turns crude oil into many products such as petrol, polythene, tar, fertilisers. Black Oil company will import the crude oil from abroad.

There are five sites available for the new Black Oil factory to be built on (see map F). The company have to decide which of the sites would be best. To do this they must look at the important location factors (shown in the box below).

Location factors for the Black Oil factory

- The oil will arrive in large oil tankers which need **deep water** for unloading.
- The refinery must be built on a large area of **flat land**.
- Many workers will be needed so the factory should be **close to a town**.
- Good **railway links** are needed for easy transport of the finished products (output).

---

2 Write out and complete the following sentence. The most suitable site for the Black Oil refinery is site _____, and the worst site is site _____.

3 Write a sentence to explain why each of the location factors shown in the photographs is important to a factory.

1 Copy the table below. Look carefully at map F. For each site (1–5), put ticks in the correct boxes to show which location factors are present. Then add the number of ticks for each site and fill in its total box. To help you, the row for site 1 has already been done.

4 Which of the location factors do you think were most important for the Daks–Simpson factory shown on page 33. Write a paragraph to explain your choices.

Site	Flat land available	Deep water available	Workforce available	Railway link available	Total
1	√	√	√		3
2					
3					
4					
5					

16 ALL CHANGE

Industrial change is not recent. British **industry** has had to cope with change for the last 200 years.

When new sources of **power** were developed, or **raw materials** ran out, many industries had to move to a new **location**. Changes in the types of transport used, such as railway engines in the 1800s and supertankers in the 1970s, also led to a change in the location of industry. As new **technology** replaced out-of-date methods, older factories closed or were replaced by modern more efficient ones.

Diagram A shows how Britain's industrial landscape has changed since the industrial revolution.

A A changing industrial landscape

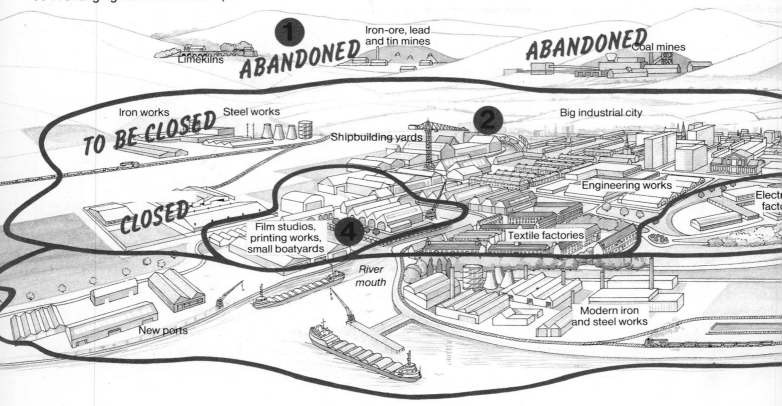

Limekilns · Iron-ore, lead and tin mines · ABANDONED · ABANDONED · Coal mines · Iron works · Steel works · Big industrial city · TO BE CLOSED · Shipbuilding yards · Engineering works · Electr factor · CLOSED · Textile factories · Film studios, printing works, small boatyards · River mouth · Modern iron and steel works · New ports

Table 1

Time	Power	Location
Early days (before 1775)	Horse, water	Upland river valleys
Industrial Revolution 1800s	Steam	Near coalfields and good trade routes
1900s	Electricity, gas, oil	Big cities, ports, old coalfields
1970 onwards	Electricity, natural gas, oil	Coasts, areas of high unemployment, new towns, industrial estates
2000	?	?

B Rehabilitated industry

Zone 1 shows the early industries of the 1700s and 1800s, now **abandoned**.

Zone 2 shows industries of the late 1800s and early 1900s which have now **closed down**.

Zone 3 shows older industries which have been moved to a new location (**relocated**).

Zone 4 shows industries which were failing but have been adapted (**rehabilitated**) to make them useful again.

Zone 5 shows very modern **high-technology** industry of the late 1900s.

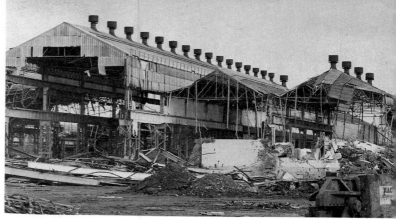

C Closed industry

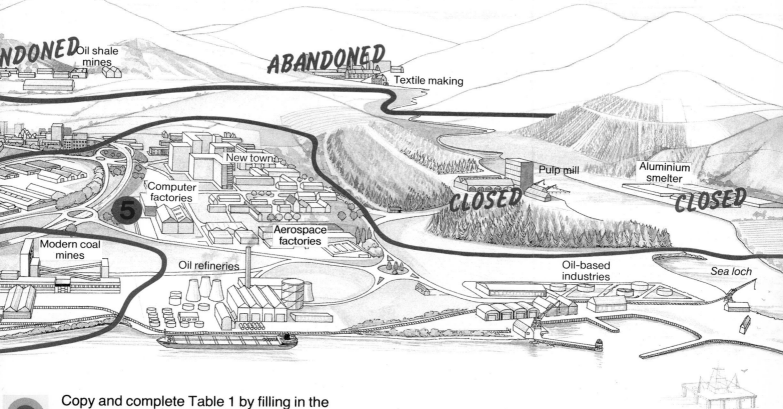

(1) Copy and complete Table 1 by filling in the changes that may take place by the year 2000. (Hint: what types of power have not been mentioned?)

(2) Look carefully at diagram A and Table 1. Explain why
(a) the early textile industry was built in highland areas;
(b) modern oil industries are built on the coast.

(3) Look carefully at diagram A. Describe the changes which have taken place in the iron and steel industry since the Industrial Revolution. What are the reasons for these changes? (Table 1 will help you.)

(4) In your own words, describe what has happened to the industrial landscape since the Industrial Revolution. Diagram A will help you. (Hint: mention spread of industry, pollution, age of building, type of industry, etc.)

(5) Name one industry which you think will grow bigger by the year 2000, and one which you think may close down altogether. Give reasons for your answers.

17 Cut and run

People remove (or **extract**) raw materials such as coal, iron-ore and oil from the earth to use in other industries, or as fuels. Coal mining, iron-ore mining and oil extraction are therefore examples of **extractive industries**. They have changed enormously over the last 200 years (see diagram A). As supplies of coal, iron-ore and oil ran out in some areas they were discovered in others, and these industries moved to new areas (**relocated**). **Abandoned** mining areas soon became **derelict** (see photo B). Tips (huge 'hills' of earth) and quarries were often left as scars on the landscape.

B Old abandoned coal mine

C Modern coal mine

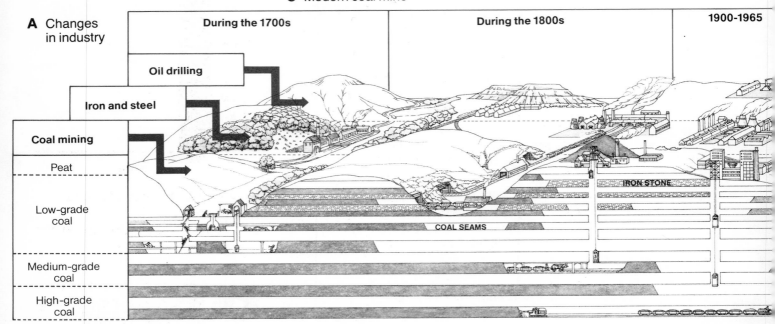

A Changes in industry

Changes in industry	1700s	1800s	1900-1965
Coal	Shallow pits, using human power, such as stair pits, bell pits	Major change to steam power; deeper pits; large tips; canal and railway transport	Very many pits change to elect
Iron-ore	River valley mills using local wood to make iron from local iron stone	Major changes as wood ran out; iron making moved to coalfields where iron-ore was also found	Iron and steel ore run out, th
Oil	Not yet important as a fuel	Oil shale mined in a few places, but supplies soon ran out or were too expensive to mine; large tips	Import of oil; so land mines

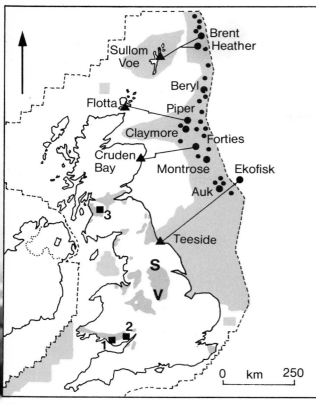

D Britain's main coalfields and oilfields

KEY

Coalfields

Modern coalfields:

S Selby

V Vale of Belvoir

Areas under oil licence

▲ Oil terminal

● • Oilfield

— Pipeline

■ Steel works:

1 Port Talbot

2 Llanwern

3 Ravenscraig

- - - Limit of British sector

0   km   250

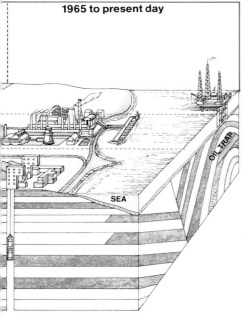

1965 to present day

1965 to present day
A few modern pits; many run-down ones close
n one large plant; as supplies of iron- move to coastal locations
North Sea oil 'boom', new technology, coastal supply and processing bases

---

New modern coal mines have now been built (see photo C and map D). The use of new technology such as coal-cutting machines, and the increasing use of oil in industry, has led to fewer coal mines and fewer jobs for miners (see Table 1).

Supplies of iron-ore in Britain have nearly run out, so the industry is located in or near to ports where iron-ore is delivered by ship from other countries. British oil refineries are also in coastal areas (see map D) since the oil now comes by pipeline from the North Sea, and by tanker from overseas.

Table 1   Coal mining in Britain

Date	Number of working mines	Number of miners
1928	1052	900 000
1950	901	691 000
1960	698	602 000
1970	282	287 000
1985	167	161 500

---

1 Neatly copy the chart below diagram A.

2 Study diagram A and explain how coal mining has changed. (Hint: size and depth of mine.)

3 Mining is an example of a primary (extractive) industry. What does this mean?

4 Look at photos B and C. Describe the differences between them.

5 Copy and complete graph E below using the information shown in column 2 of Table 1.

E

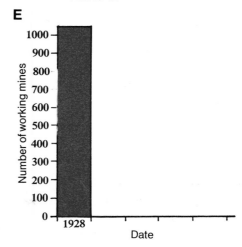

(a) What has happened to coal mining since 1928?

(b) Give two reasons to explain this change.

6 Look carefully at map D. Make a list of the following:
(a) modern coalfields;
(b) coastal steel works;
(c) modern oil fields.

7 Industrial change always creates some problems. Describe some of the problems that you think might arise in an area where mines have been abandoned.

39

18 UP FOR SALE!

Old industries become less important (**decline**) as time passes, and new industries grow to take their place. This type of change has caused even more problems than those found in extractive industries (see unit 17).

Manufacturing industries such as textiles, shipbuilding, steel, engineering and, lately, motor vehicles have become less important (see Table 1). These **traditional industries** have suffered most. Many factories have closed down (see photo A) with thousands of jobs lost. There has been a shift away from the old industrial areas (map B) towards new industrial areas, especially to the south-east of England, around London (map A, page 42). Here, newer industries such as electronics have become more important.

This shift has created many problems for old industrial areas such as Salford in Greater Manchester (photo C and sketch D). These areas were once Britain's great workshops. Today, they are in decline and are areas of high unemployment. Not enough new industries are being set up in areas like these, and most of the new factories only employ a small number of people.

A Consett steelworks after closure

B Britain's old industrial areas

KEY

■ Older industrial areas

1 Numbers refer to 'Up for sale' industries in Table 1

Table 1 'Up for sale'

Industry	Number on map B	Large yard closed	Year closed	Jobs lost
Steel	1	Ebbw Vale, wales	1978	7 500
	2	Consett, North-east	1980	5 000
	3	Corby, East Midlands	1980	8 000
	4	Shotton, North-west	1981	9 000
	5	Gartcosh, Scotland	1986	800
	6	Ravenscraig, Scotland	?	?
Vehicles	7	Canley, Midlands	1980	6 000
	8	Abingdon, South Midlands	1980	4 000
	9	Linwood, Scotland	1981	8 000
	10	Speke, North-west	1982	6 000
	11	Solihull, Midlands	1982	9 000
	12	Bathgate, Scotland	1986	800
Engineering	13	Singers, Scotland	1979	3 000
	14	Herberts, Midlands	1982	5 000
Aluminium smelting	15	Invergordon, Scotland	1982	2 300
Pulp-making	16	Corpach, Scotland	1980	450

C An old industrial area

D Sketch of photo C

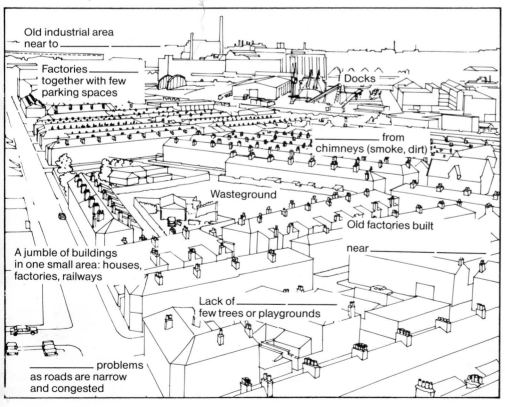

Old industrial area near to _____

Factories _____ together with few parking spaces

Docks

_____ from chimneys (smoke, dirt)

Wasteground

Old factories built

A jumble of buildings in one small area: houses, factories, railways

near _____

Lack of _____ few trees or playgrounds

_____ problems as roads are narrow and congested

---

1 Steel-making is an example of a manufacturing industry. Explain what this means.

2 Make a list of the traditional industries mentioned on these pages.

3 Compare map B with map D on page 39. Where are most of the traditional industries located? Why is this?

4 Study Table 1 carefully. Use the information shown to work out
(a) the total number of jobs lost in steel;
(b) the total number of jobs lost in motor vehicles;
(c) the worst year for Britain's traditional industries.

5 Study sketch D. Write the title 'An old industrial area'. Below it copy and complete the labels from the sketch by using words from the box below.

pollution,	city centre,
housing area,	cramped,
open spaces,	transport

6 Copy and complete the following information. Choose words from the box below.

Many _____ industries, such as textiles and steel, have closed down. There has been a _____ away from the _____ industrial areas towards the _____ industrial areas, especially in _____ Britain. _____ is now a big problem in old industrial areas.

Choose from: south-east, new, shift, old, traditional, unemployment

19 A way ahead

Map A shows some of Britain's new industrial areas. More than half of the new industries are in south-east England. New industries often choose to set up in places where there was very little industry before. A good example is the area around Cambridge. It houses over 300 **high-technology industries** which specialise in **micro-electronics**.

New industrial areas such as Cambridge Science Park (map B) are very different from old industrial areas. They are carefully planned and built in a modern clean environment, often on the edge of cities beside good transport routes (see chart C).

A large variety of products are made by 'hi-tech' industries. These range from home computers to laser and aerospace equipment (see chart D). Most of these products can be easily transported long distances because they are small and light.

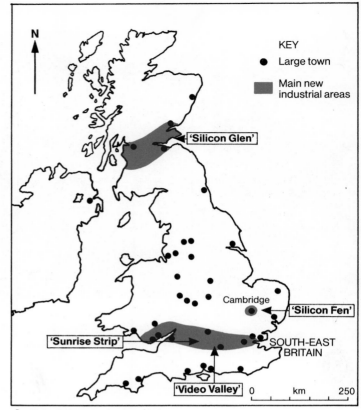

KEY
● Large town
■ Main new industrial areas

'Silicon Glen'
Cambridge
'Silicon Fen'
'Sunrise Strip'
SOUTH-EAST BRITAIN
'Video Valley'
0   km   250

A Britain's new industrial areas

C Location factors for new industries

- Electricity must be easily available (this means almost anywhere!)
- Land must be cheap and plentiful
- Big towns should be nearby to provide markets for goods (areas such as south-east England, central Scotland)
- Good communications are needed (roads, railway links)
- Universities where high-technology research is being carried out should be nearby.

B Cambridge Science Park

N
0   m   250
A45 Northern by-pass
Space for more industrial units
Napp Laboratories
A1309

KEY
▨ Farmland
■ Industrial units and factories
▤ Lake
▦ Grass and trees
╫ Railway
— Road

Housing

D High-technology products

Computer systems
Micro-computers
Software
Medicines and vaccines
Laser technology
Aerospace instruments
Radio-telephone systems

Fibre-optic cables
Weather instruments
Synthetic foodstuffs
Metal-alloy research
Optical products
Energy conservation products

1. Make a list of the new industries mentioned on these pages.

2. Write a sentence for each of the following to explain why new industrial areas
(a) do not need to be beside coalfields;
(b) need lots of flat land to build on;
(c) do not need to be built in the middle of housing areas.

3. Study sketch F. Write the title 'A new industrial area'. Below it copy and complete the labels from the sketch by using words from the box below.

away from,	pollution,	clean,
new,	roads,	workers,	grass,
on the edge,	car parks	

4. Look carefully at map B.
(a) Why are the 'A' roads important in the location of Cambridge Science Park?
(b) Why have the designers of Cambridge Science Park taken the trouble to include lakes, ponds and trees in the layout?

5. Look carefully at photo E above and photo C on page 41. Copy and complete Table 1 choosing the correct phrases from the list below.

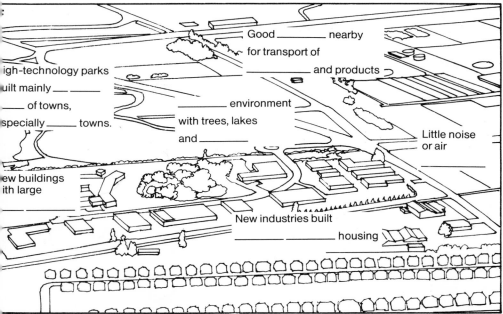

Good _____ nearby for transport of _____ and products

High-technology parks built mainly _____ _____ of towns, specially _____ towns.

_____ environment with trees, lakes and _____

Little noise or air

New buildings with large _____

New industries built _____ _____ housing

Table 1 Old and new industrial areas

	Old industrial areas	New industrial areas
1. Location of industry		
2. Location of houses		
3. Layout of houses		
4. Height of factories		
5. Layout of factories		
6. Type of power used		
7. Main type of transport		
8. Products		
9. Surrounding area		
10. Pollution		

1. On outskirts/Near city centre
2. Far from factories/Near to factories
3. Cramped together/Well spread out
4. 1–2 storeys/Multi-storeys
5. Cramped together/Well spaced out
6. Was coal, now electricity/Electricity
7. Roads/Canals, railways
8. Mostly heavy weight/Mostly light weight
9. Run-down/Landscaped, well planned
10. Clean air/Smoke, waste dumping

20 Spread the Risk

Industrial change sometimes means factories or companies closing down or moving to a new area. If companies make only one product, it is more likely that they will have to close if sales of their product drop.

To avoid this danger, some companies **diversify**. This means that they make different products. They may do this by taking over other companies. If one product fails, the company can survive on its other products.

Arthur Guinness and Sons PLC is a good example of a company which has diversified its interests. The company was founded in 1759 in the Republic of Ireland, where it brewed its dark beer. In 1936 it opened a London brewery, and since 1963 Guinness has spread to over 140 countries around the world (see map A).

Guinness is made from four ingredients (see diagram B) in a process which has changed little since 1759. The early company decided to make only one product, then known as 'porter'. It became so successful that, by 1868, the Guinness company had become the world's largest brewing company. Today it is Europe's largest, and Guinness remains a very popular beer.

A Guinness around the world

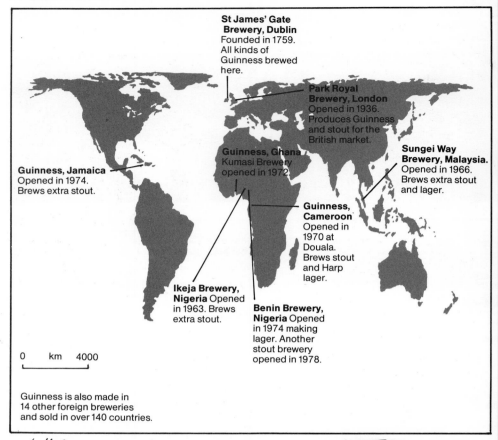

St James' Gate Brewery, Dublin Founded in 1759. All kinds of Guinness brewed here.

Park Royal Brewery, London Opened in 1936. Produces Guinness and stout for the British market.

Guinness, Jamaica Opened in 1974. Brews extra stout.

Guinness, Ghana Kumasi Brewery opened in 1972.

Sungei Way Brewery, Malaysia. Opened in 1966. Brews extra stout and lager.

Guinness, Cameroon Opened in 1970 at Douala. Brews stout and Harp lager.

Ikeja Brewery, Nigeria Opened in 1963. Brews extra stout.

Benin Brewery, Nigeria Opened in 1974 making lager. Another stout brewery opened in 1978.

0   km   4000

Guinness is also made in 14 other foreign breweries and sold in over 140 countries.

Barley

Water

Hops

Yeast

Sugar

B

Bottles and cans

'Draught' barrels

Arthur Guinness and Sons PLC now employ over 12 000 people. The company has diversified from making only one product (Guinness) to many different interests (see diagram D). As a result, the company has become stronger and is less likely to suffer from the problems of industrial change.

C A Guinness advert.
The Guinness word and harp device are registered trade mark

D Guinness spread the risk

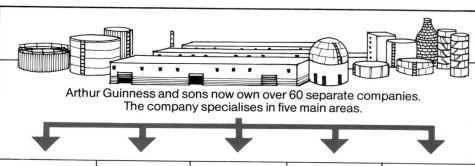

Arthur Guinness and sons now own over 60 separate companies.
The company specialises in five main areas.

BREWING	DISTILLING	PUBLISHING	HEALTH AND FITNESS	SHOPS AND STORES
Brewing of Guinness, Harp lager, stout and non-alcoholic lager. Many public houses and hotels. Also tin-can and barrel-making factories.	In 1985, the Guinness Co. bought Bells Scotch Whisky Ltd. It also took over Bells hotel chain called 'Gleneagles'. In 1986, the company diversified by taking over the Distillers Company.	World famous publications such as the *Guinness Book of Records* and *Guinness Book of Hit Singles*. Sales of over 4 millions each year. Sold in 146 countries and 22 languages.	Guinness took over two health companies in 1984 and 1985. This gives the company health spas in Tring and Edinburgh and 'Nature's Best Health Products Ltd'.	Guiness now own over 650 shops after buying 'Martins' newsagent chain and 'Neighbourhood Stores'.
Type of industry	**Type of industry**	**Type of industry**	**Type of industry**	**Type of industry**

1 What happens when a company diversifies? Write a few sentences to explain.

2 Study diagram B. Then answer the following questions.
(a) Which five ingredients are used to brew Guinness?
(b) What is draught beer?
(c) Brewing is a manufacturing industry. Explain what this means.

3 Why do you think Guinness is made in countries throughout the world rather than making it all in Ireland? (Hint: time, transport cost.)

4 Where would be the best sort of place to build a Guinness factory in Britain? Make a list of all the important factors a company would need to take into account before deciding. (Hint: think about raw materials, workers, power, transport, selling markets, etc.)

5 Neatly copy diagram D. Complete the 'type of industry' boxes by choosing from: primary, manufacturing, services. (Hint: there may be more than one type for each box.)

Investigating
Have any companies in your local area diversified? Write to some large companies to find out.

21 A Fresh Start

A London's city-centre docklands

London's docklands (see photo A) made the city into an important centre for trade and manufacturing industries. At one time the 2000 hectares of dockland in the city centre had nine working docks (see map B), many shipyards, factories, warehouses and dockers' houses.

By the end of the 1950s, however, modern ships were too big for the central part of the River Thames. Industries moved east to the mouth of the river and new docks opened there. Warehouses and shipyards in the old docklands closed, and thousands of jobs were lost (see diagram C). By 1980, London's old docklands were in a sad state.

 What happened to London's docklands between 1950 and 1980? Why?

C London's closing docklands

B London's quay to success

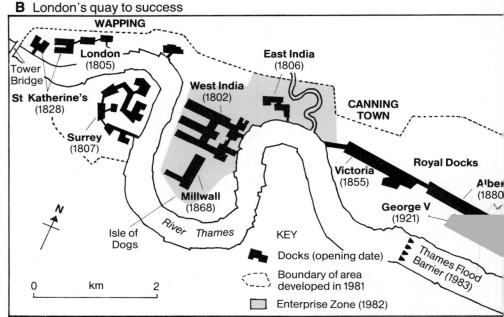

WAPPING

London (1805)

Tower Bridge

St Katherine's (1828)

East India (1806)

West India (1802)

CANNING TOWN

Surrey (1807)

Royal Docks

Victoria (1855)

Albert (1880)

Millwall (1868)

George V (1921)

Isle of Dogs

River Thames

Thames Flood Barrier (1983)

KEY

Docks (opening date)

Boundary of area developed in 1981

Enterprise Zone (1982)

0   km   2

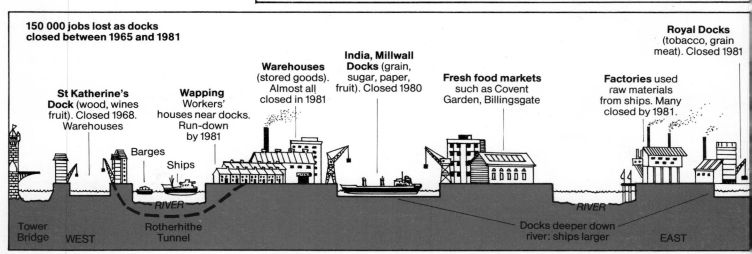

150 000 jobs lost as docks closed between 1965 and 1981

Royal Docks (tobacco, grain, meat). Closed 1981

St Katherine's Dock (wood, wines, fruit). Closed 1968. Warehouses

Wapping Workers' houses near docks. Run-down by 1981

Warehouses (stored goods). Almost all closed in 1981

India, Millwall Docks (grain, sugar, paper, fruit). Closed 1980

Fresh food markets such as Covent Garden, Billingsgate

Factories used raw materials from ships. Many closed by 1981.

Barges

Ships

RIVER

Tower Bridge   WEST   Rotherhithe Tunnel

Docks deeper down river: ships larger

RIVER

EAST

D St Katherine's Dock

A fresh start was needed. It came when the government set up the London Docklands Development Corporation (L.D.D.C.) in 1981. Its job was to **rehabilitate** the old docklands (adapt them to suit new needs). The fresh start seems to have been successful (see photo D and diagram E). So far the L.D.D.C. has

- improved roads and railways in almost all parts of the docklands, especially in the Isle of Dogs;
- brought in over 100 new industries to occupy modernised warehouses, providing over 6000 new jobs;
- modernised docks such as St Katherine's (see photo D);
- improved the standard of housing, built over 6000 new houses and tidied up the look of the area;
- set up the Isle of Dogs **Enterprise Zone** which helps new industries by giving grants and charging no rates on buildings.

The rehabilitation of areas such as London's docklands is another answer to the problems of change in Britain's old industrial areas.

E A fresh start for London's old docklands

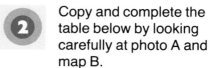

Copy and complete the table below by looking carefully at photo A and map B.

Area	Name of dock area
a	
b	
c	
d	
e	

3 What happens when an industrial area is rehabilitated? Write a few sentences to explain.

4 Copy diagram E. Look carefully at diagram C, then use it (and map B) to help you copy and complete the following key for your diagram.

1. ___ _____ Dock now has offices, hotels, a world trade centre and 1700 new jobs.
2. New _____ built.
3. Rehabilitated _____.
4. Some _____ filled in.
5. New roads and railways.
6. New Billingsgate fish _____.
7. Thames F _____ Barrier (1983).
8. Old _____ split up into small modern industrial units.
9. _____ used for water sports.
10. Docklands airport planned.

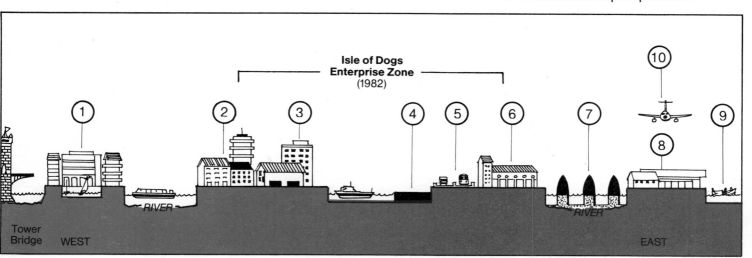

22 ALTERED IMAGES

Glasgow's image today is very different from its image ten or twenty years ago. At that time Glasgow was seen as an old industrial city where pollution, slums and poverty were among the worst in Britain. Today Glasgow puts forward an image of a bright, clean and interesting city with friendly people and lots to see and do.

Has Glasgow really changed so much? Yes . . . and No!

Glasgow still has high unemployment and some of the poorest housing in Britain. But new exciting things are happening, especially in the city centre. Glasgow is changing from an industrial city to a commercial (business) one.

The information and pictures on these pages show some of the changes that are taking place in Glasgow today.

Abandoned dock areas had been run down and unused for many years. Custom House Quay is an example of how these areas can be made attractive.

Many new office blocks and luxury hotels are being built in the city.

The Scottish Exhibition Centre, built a little west of the city centre, is a good example of the way Glasgow is looking ahead. The Exhibition Centre is built on old dockland and is near main roads and the airport. It has good conference facilities which attract business people.

There have been schemes to improve the old docklands and areas of old industry and housing.

The Glasgow East Area Renewal (G.E.A.R.) project has been given government money to improve the worst parts of the city. New jobs have been created, new houses and factories built, and old houses have been improved.

Many old buildings have been given a 'facelift'. Stone cleaning and modernisation are examples of what can be done.

There is excellent access to the city centre by road, rail and air (Glasgow Airport is only ten minutes away from the city centre by car).

In the city centre, some streets have been closed to traffic and are used by pedestrians only. New shopping centres have opened, and car parking has been restricted.

There is a new urban motorway which does not go round the city centre as usually happens, but right through the middle.

The city puts forward its new image with cheerful advertising: the 'Glasgow's Miles Better' slogan says it all.

GLASGOW'S MILES BETTER
A GREAT PLACE TO LIVE AND WORK

1 Using the photographs and boxed information on these pages, make a list of the modern city features found in Glasgow today.

2 In what ways is modern Glasgow like your nearest city?

3 Which of the modern city features shown or described are
(a) completely new,
(b) modernised?

4 Which features of modern Glasgow do you find attractive? What makes them appeal to you?

5 Which types of people do you think Glasgow is trying to attract with its advertised new image?

6 Why do you think the docks in central Glasgow are no longer used for shipping?

23 Transclyde

Until about 1870, the only way that most people could travel about a city was on foot. The invention of the tram (photo A) meant that cheap transport of lots of people was possible in the city centre.

At about the same time, large railway stations were opening in cities all over Britain. For the first time it became possible to live outside a city and travel into work every day.

Map B shows four main railway stations in Glasgow in 1885. **Commuters** travelled into Glasgow from all directions (photo C). Transport in Glasgow city centre was further improved when an underground railway was built in 1897, motorbuses appeared in 1924, and trolley buses in 1949.

All these types of transport worked separately. In time, some, such as trams and trolley buses, went out of use. In the 1970s, transport planners began to develop an **integrated transport plan** which linked all available types of transport as efficiently as possible. The Transclyde network was born. Rail, bus and underground services were modernised, some railway stations were closed, others opened, and bus and train interchanges were started (map E). This meant that bus stops were placed next to underground and railway stations, so that passengers could change easily from one form of transport to another without delay.

The increasing use of cars meant that city-centre streets were congested with traffic, and car parks in the centre of the city were full (photo F). To avoid this, a 'park and ride' scheme was introduced. Car parks were built beside underground stations at the edge of the city so that commuters could leave their cars there and use public transport.

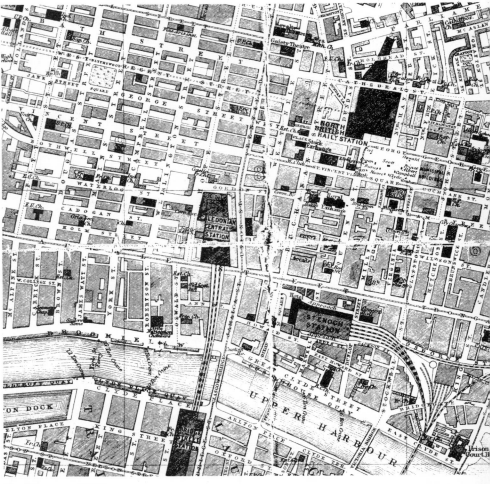

B Glasgow 1885

A Tram in Glasgow, 1950s

C Steam train and commuters in Glasgow Central Station, 1950s

1 What is a commuter?

2 For what reasons might someone travel to the city
(a) daily,
(b) weekly,
(c) in the evenings,
(d) occasionally?

3 Use the photographs and maps to help you describe the changes which have taken place in transport in Glasgow since 1885. Mention car parks, bus stations, underground stations, railway stations, the road system.

4 Using the Transclyde network diagram (E) describe your journey to work in Glasgow's city centre from Motherwell.

5 What changes have made it possible for people to live further away from their work?

6 (a) What does the 'park and ride' sign mean? (See map E.)
(b) Why do you think car parking is free for underground passengers in some cities?

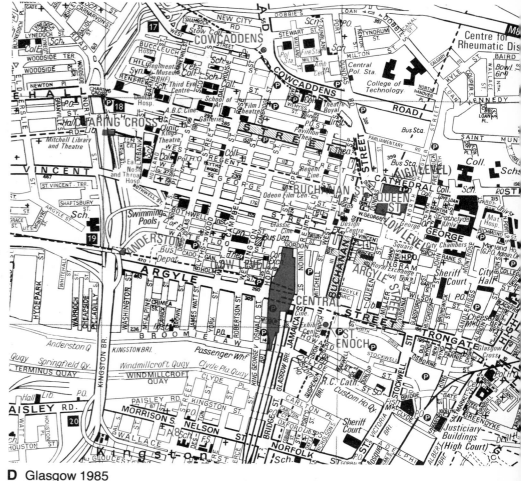

D Glasgow 1985
Reproduced with the permission of Geographia Ltd, London, Crown Copyright reserved

F A city-centre car park

E Part of the Transclyde rail network

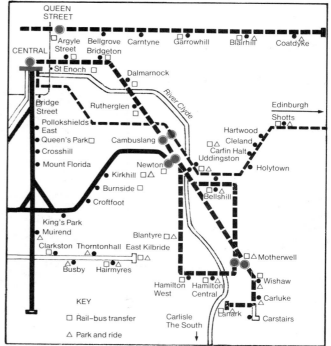

KEY
□ Rail–bus transfer
△ Park and ride

Pollution gets up people's noses. It spoils and destroys the environment (see photo A), it chokes rivers (see photo B), spoils beaches and kills wildlife. We are all 'waste watching' on a diet of rubbish.

As towns, cities and industries have grown and spread over Britain's landscape, so have the problems of pollution. Diagram C shows many ways in which the environment can be polluted. Some types of pollution, such as chimney smoke, last for a short time (or **short span**). Other types of pollution, such as radio-activity or coal tips, last for a very long time (or **long span**). Pollution can be sorted into five main types.

- **Air** pollution, including acid rain.
- **Water** pollution of seas, rivers, lakes and lochs.
- **Noise** pollution: very loud noise can damage hearing.
- **Land or soil** pollution, including soil poisoning, quarries and tips.
- **Visual** pollution, when the landscape becomes unsightly.

In some places, improving the environment by getting rid of pollution is now a long overdue job.

1 What is pollution? Write a sentence to explain.

2 Look carefully at diagram C. Then copy and complete the table below by writing as many examples of each type of pollution as you can find in the diagram.

Type of pollution	Examples
Air	
Water	
Noise	
Land or soil	
Visual	

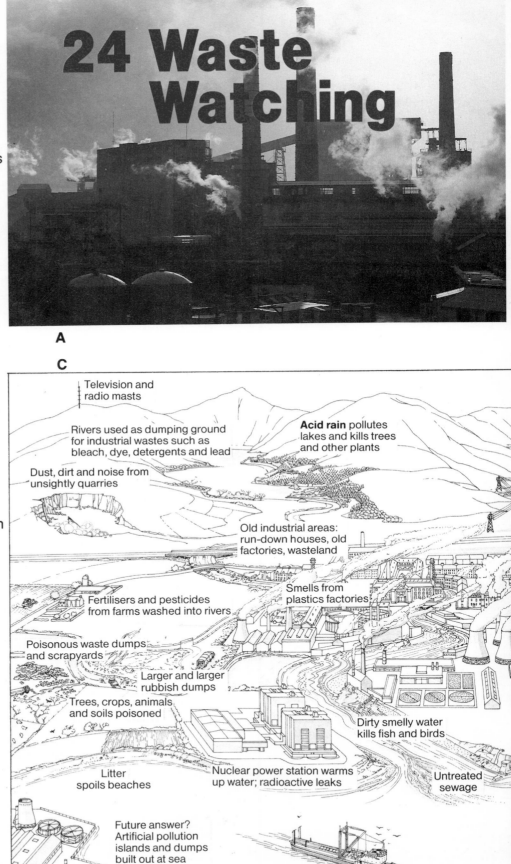

24 Waste Watching

A

C

Television and radio masts

Rivers used as dumping ground for industrial wastes such as bleach, dye, detergents and lead

Acid rain pollutes lakes and kills trees and other plants

Dust, dirt and noise from unsightly quarries

Old industrial areas: run-down houses, old factories, wasteland

Smells from plastics factories

Fertilisers and pesticides from farms washed into rivers

Poisonous waste dumps and scrapyards

Larger and larger rubbish dumps

Trees, crops, animals and soils poisoned

Dirty smelly water kills fish and birds

Litter spoils beaches

Nuclear power station warms up water; radioactive leaks

Untreated sewage

Future answer? Artificial pollution islands and dumps built out at sea

B

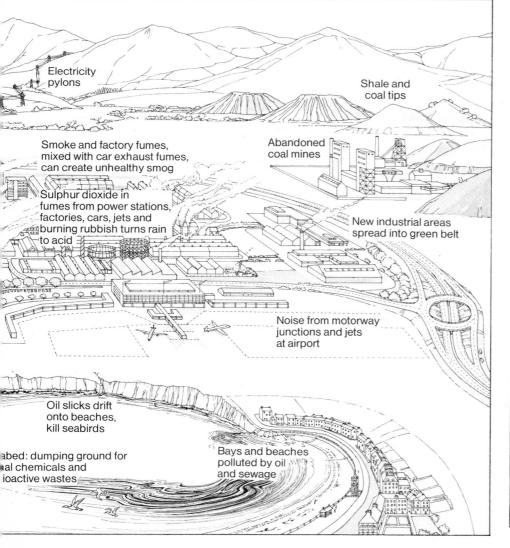

Electricity pylons

Shale and coal tips

Smoke and factory fumes, mixed with car exhaust fumes, can create unhealthy smog

Abandoned coal mines

Sulphur dioxide in fumes from power stations, factories, cars, jets and burning rubbish turns rain to acid

New industrial areas spread into green belt

Noise from motorway junctions and jets at airport

Oil slicks drift onto beaches, kill seabirds

...abed: dumping ground for ...al chemicals and ...ioactive wastes

Bays and beaches polluted by oil and sewage

3 For each of the examples of pollution that you have listed in your table, say whether it is short span (lasts a short time only) or long span (lasts a long time).

4 What do you think each of the following people would have to say about the pollution shown in photos A and B?
Photo A: farmer, factory manager.
Photo B: factory supervisor, fisherman.
Discuss this in a group first. Then write a couple of sentences giving each person's point of view.

5 The environment can be improved in many ways. Copy and complete the following table by using the phrases in the box below.

Pollution	Improvements
Smoke and factory fumes	
Old quarries and coal tips	
Industrial waste in rivers	/
Old run-down industrial areas	
Rubbish tips	

Choose from:
Using laws to make factories clean up waste
Changing old houses and industries to suit new uses
Using bottle banks and recycling metal and paper
Using the Clean Air Act (1956)
Removing, filling in and landscaping

25 Gloomy Prospects

'Have you heard about the artist who has become unemployed?
 Now he's drawing the dole.'

Jokes such as this are no longer funny. Today over 13% of Britain's workforce (one in every seven) are out of work. The local Jobcentre (photo A) is one of the busiest 'shops' in town. Young people are particularly badly hit. Unemployment has risen steadily over the last ten years. Fewer and fewer young people are able to find jobs or apprenticeships. Many are trapped in the vicious circle of 'no experience – no job' (diagram B) and some may never work in their lifetime.

Not all parts of Britain suffer equally from high unemployment (see map C). Some areas, such as south-east England, have much less unemployment than average. Other areas, such as Northern Ireland, Scotland, Wales and north and west England are worse off than average. Industrial change has been the reason for the very high unemployment in these areas. Older industries such as textiles and shipbuilding have become less important and have died out in many areas. Newer industries such as micro-electronics have been set up (especially in the south-east) but these do not often solve the problems of replacing old industries because they are set up in different areas. Also, they employ fewer people because they use modern technology. Since the 1970s more jobs have been lost than new ones have been created.

A

B Experienced young person wanted

① Young person. I'd like to apply for the job please

② Employer. What experience do you have?

③ Young person. None-this would be my first job.

④ Employer. Sorry- no experience – no job.

⑤ Young person. But how can I get experience?

⑥ Employer. Easy-get yourself a job!

C Unemployment in Britain

1 Scotland 15.0%	7 Yorkshire & Humberside 14.4%
2 Northern Ireland 21.0%	8 East Midlands 12.3%
3 Wales 16.4%	9 East Anglia 10.5%
4 North England 18.1%	10 South-east England 9.7%
5 North-west England 15.8%	11 South-west England 11.8%
6 West Midlands 14.9%	

KEY

Very much above average

Well above average

Above average

Below average

Well below average

N

Average: 13.1% of the workforce is unemployed

0   km   250

Reasons to be cheerful?

All is not lost, however! Today in Britain, more people work in **service industries** – for example in hotels, schools, hospitals, entertainment and leisure – than in **manufacturing industries** (see diagram D). As the standard of living has risen, people have more free time and can afford more and better services. The continuing growth in service industries is one reason to be cheerful as this provides people with new jobs.

When the huge Linwood car assembly plant in Paisley closed in 1981, 5000 people lost their jobs. The Paisley area was hit by an unemployment crisis. Many people left in search of work elsewhere and the huge factory stood empty, up for sale. In 1984 the Portal Company took over the factory and split it up into smaller units (see photo E). Today the Junction 29 site of the old Linwood plant houses eight companies employing up to 2000 people. Creating new jobs in re-opened factories may be a way forward for areas with unemployment problems.

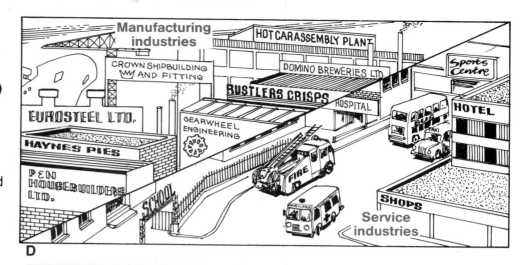

D

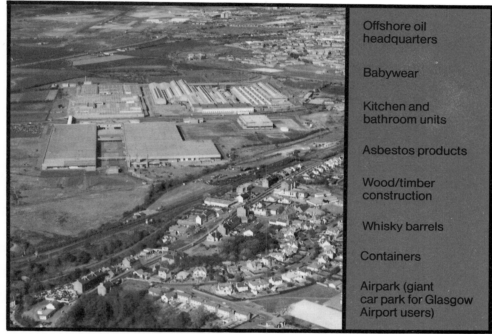

Offshore oil headquarters

Babywear

Kitchen and bathroom units

Asbestos products

Wood/timber construction

Whisky barrels

Containers

Airpark (giant car park for Glasgow Airport users)

E Linwood site today (Junction 29)

1 Use map C and its key to help you copy and complete the table below. Put the regions in order: the region with the highest percentage unemployed first, and the region with the lowest percentage unemployed last.

Region	Percentage unemployed

2 Read the text carefully again. What are the main reasons for
(a) high unemployment in Wales, Scotland, north and west England?
(b) low unemployment in south-east England?
Write a few sentences to explain.

3 What do we mean by 'service industries'? Make a list of more than ten (see diagram D).

4 Write a sentence to explain how each of the following might affect unemployment:
(a) a shorter working week;
(b) growth of more services;
(c) job-sharing schemes;
(d) more small businesses;
(e) use of more modern technology such as robots, micro-electronics and lasers.

26 A helping hand

One of the most important factors in the location of British industry has been the helping hand given by the government. The government encourages industry to move to the areas that need new jobs most: those with high unemployment. These parts of Britain are called **Development Areas**. Today, they are areas where old industry is dying out, or places which are isolated (see map A). **Intermediate Areas** are also given some help but not as much as Development Areas.

The government offers a wide range of **incentives** (or benefits) to industries if they decide to set up in Development Areas or Intermediate Areas (see Table 1). Many other types of help are also available, such as

- **retraining** schemes for workers;
- building of '**advance factories**' (photo B) which are ready-made for firms to move into;
- new towns creating jobs in new factories;
- job creation and youth training schemes;
- improvements in transport links (photo C, page 61);
- **Enterprise Zones** (map D and sketch E, page 61): these are towns which are especially badly hit by unemployment and get even more help from the government than Development Areas (see Table 1).

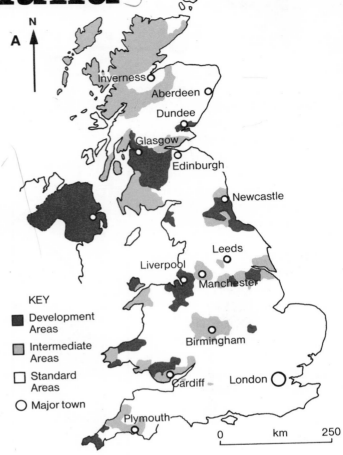

KEY
- Development Areas
- Intermediate Areas
- Standard Areas
- ○ Major town

B Advance factory units

Although governments have been helping industry in Development Areas for over 40 years, some people think that Britain still has a poor unemployed north and a more wealthy employed south. But, without the government's helping hand, the problems in the Development Areas would be much worse.

Table 1 Helping hands: government aid for industry

Incentive	Intermediate Areas	Development Areas	Enterprise Zones
Government-built factories	Up to 2 years rent free	Up to 2 years rent free	Up to 10 years rent free
Building grants (paid by government)	15%	15%	Up to 100% grants
Government loans	No	Yes	Yes
Plant and machinery grants	Up to 15%	15%	100%
Training grants	Free training at a government skills centre		Free at skills centre
Grants for removal costs	Up to 80%	Up to 80%	Up to 100%
Help for transferred workers	Free fares, lodgings and help with removal costs		
Tax allowances	Less tax to be paid and rates reduced for up to 5 years		Up to 10 years free tax

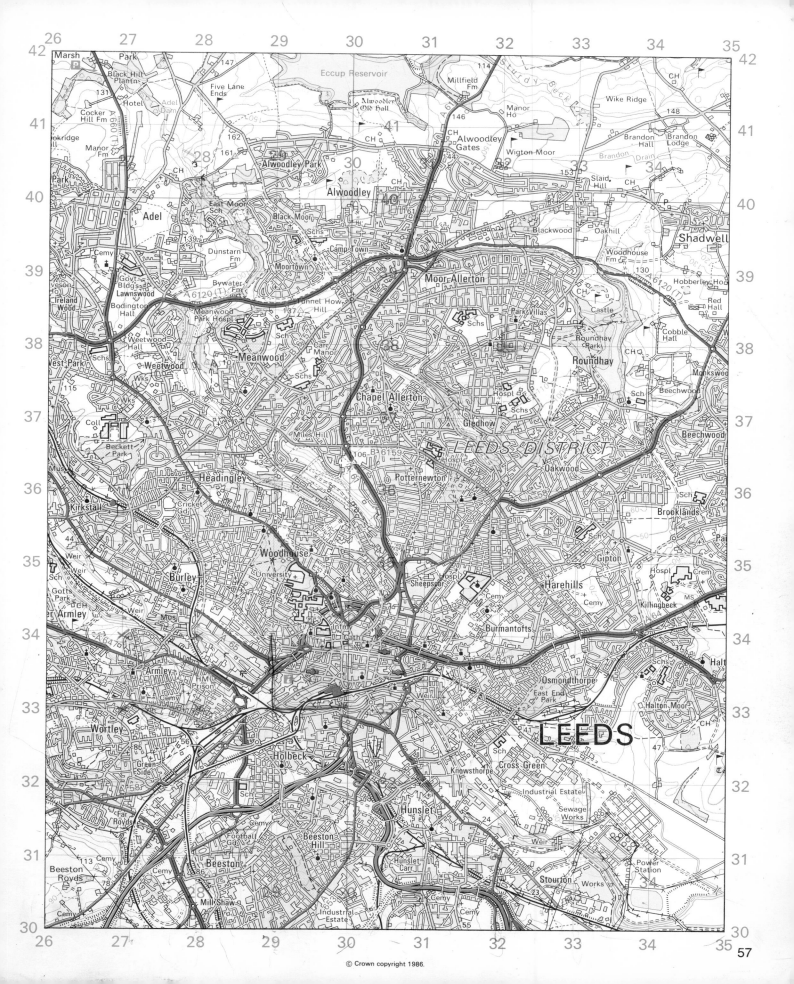

© Crown copyright 1986.

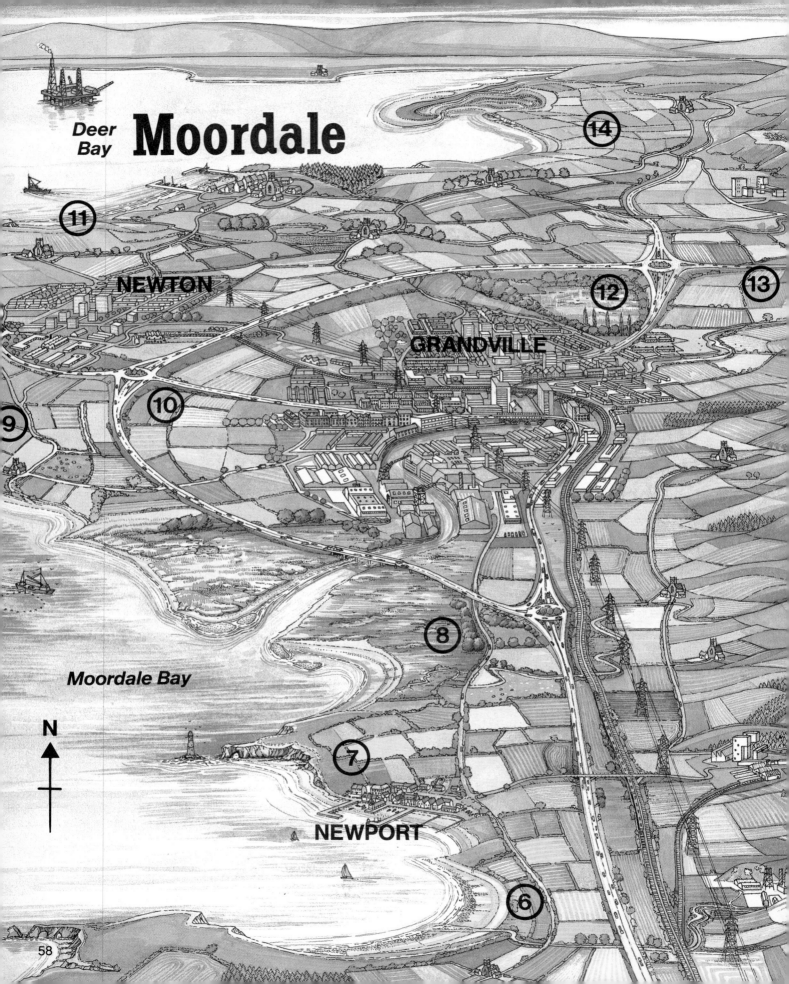

Deer
Bay

Moordale

NEWTON

GRANDVILLE

Moordale Bay

N

NEWPORT

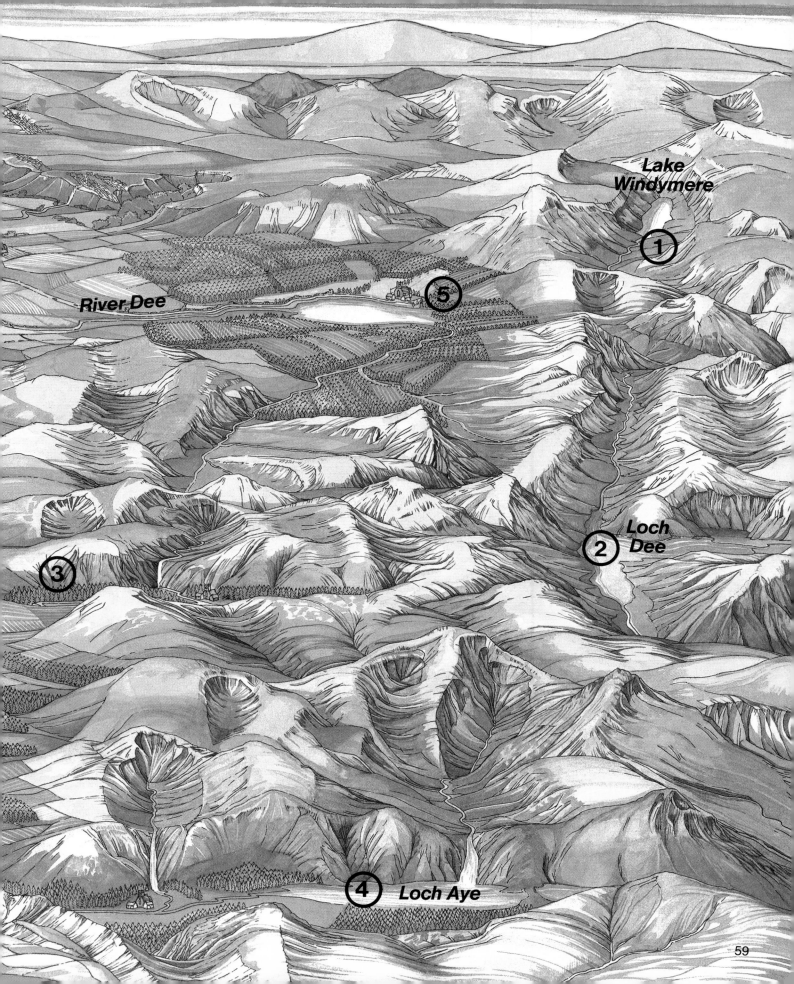

Lake Windymere

River Dee

Loch Dee

Loch Aye

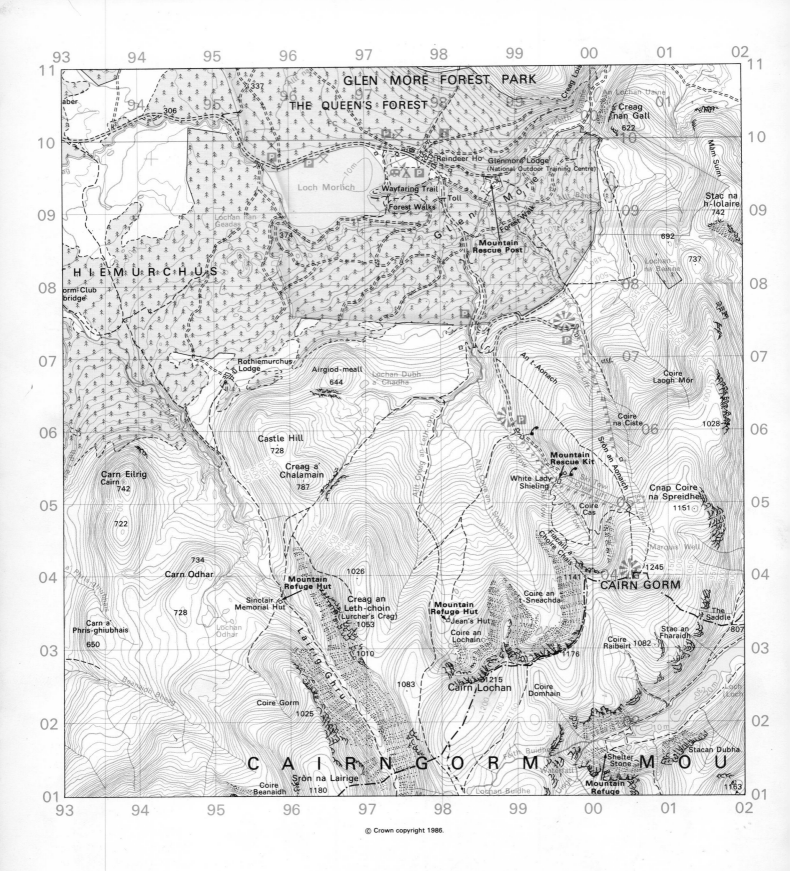

GLEN MORE FOREST PARK

THE QUEEN'S FOREST

Loch Morlich

ROTHIEMURCHUS

Reindeer Ho
Glenmore Lodge
(National Outdoor Training Centre)

Wayfaring Trail
Toll
Forest Walks

Mountain
Rescue Post

Creag
nan Gall
622

Stac na
h-Iolaire
742

692

737

Rothiemurchus
Lodge

Airgiod-meall
644

Lochan Dubh
a' Chadha

An t-Aonach

Coire
Laogh Mór

Coire
na Ciste

1028

Castle Hill
728

Carn Eilrig
Cairn
742

Creag a'
Chalamain
787

Mountain
Rescue Kit

White Lady
Shieling

Coire
Cas

Cnap Coire
na Spreidhe
1151

722

Marquis Well

734

Carn Odhar

1026

1141

1245

CAIRN GORM

Carn a'
Phris-ghiubhais
650

728

Mountain
Refuge Hut

Sinclair
Memorial Hut

Lochan
Odhar

Creag an
Leth-choin
(Lurcher's Crag)
1053

Mountain
Refuge Hut

Jean's Hut

Coire an
t-Sneachda

Coire an
Lochain

The
Saddle
807

Stac an
Fharaidh

Coire
Raibeirt 1082

1010

1176

1215

Lairig Ghru

1083

Cairn Lochan

Coire
Domhain

Coire Gorm
1025

C A I R N G O R M   M O U

Shelter
Stone

Stacan Dubha

Coire
Beanaidh

Sròn na Lairige
1180

Mountain
Refuge

1163

© Crown copyright 1986.

60

1 Look carefully at map A and compare it with the map of unemployment (map C, page 54). How are they similar?

2 What is a Development Area? Give some examples in your answer.

C 'Spaghetti Junction'

E Clydebank Enterprise Zone

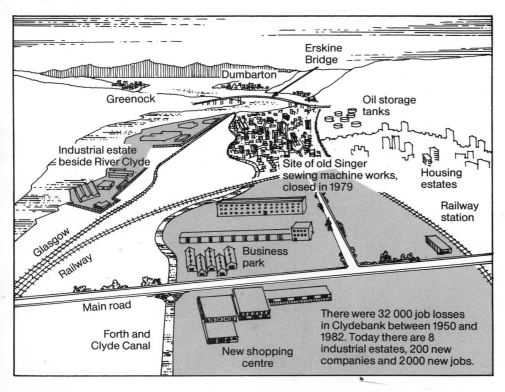

Erskine Bridge

Dumbarton

Greenock

Oil storage tanks

Industrial estate beside River Clyde

Site of old Singer sewing machine works, closed in 1979

Housing estates

Railway station

Glasgow Railway

Business park

Main road

Forth and Clyde Canal

New shopping centre

There were 32 000 job losses in Clydebank between 1950 and 1982. Today there are 8 industrial estates, 200 new companies and 2000 new jobs.

N

Invergordon

Dundee

Clydebank

Belfast

Gateshead
Hartlepool

Speke

Wakefield

Salford

Dudley

Swansea

Corby

Isle of Dogs, London

0   km   250

D Enterprise Zones, 1985

3 How does the government persuade companies to set up business in Development Areas? Write a few sentences to explain.

4 Why are there no Development Areas in south-east Britain?

5 If you were to set up a large factory, which of the following three types of areas – Development Area, Intermediate Area or Enterprise Zone – would you choose? (See Table 1). Give reasons for your answer.

6 Why do you think many companies stay where they are rather than move to a Development Area?

7 Use Table 1 and sketch E to help you make a list of advantages that a new company would have if it set up in Clydebank Enterprise Zone. (Hint: Incentives, transport links, workforce, pleasant location, etc.)

27 RIVER DEEP, MOUNTAIN HIGH

A Europe's wild landscapes

If you took a journey across Europe you would pass through many different landscapes. There are industrial landscapes such as the Ruhr (West Germany) and the coalfield areas of northern France. Then there are the great cities such as Paris, London, Madrid and Rome, which are the homes and workplaces of millions of people.

As well as these **urban landscapes**, Europe has vast areas of **rural landscape**. Over the centuries, people have drained marshes, and felled forests to clear the land for farming.

All of these urban and rural landscapes have been made by people. But there are still parts of Europe which have changed very little or not at all. These are the areas of **wild landscape** such as high mountains, long deep rivers, volcanoes, cold deserts and rocky or marshy plains. Some of these wild areas are marked on map A and shown in photographs B–G.

Most of Europe's wild areas are difficult places to live, but some are more attractive and, with the help of modern machinery and ideas, could be changed or 'tamed'.

The Camargue area of France is an example (see photo D). This is a marshy area of grassland on the delta of the Rhone river, famous for its wild horses and other wildlife. Some farmers want to drain the area to farm the rich soils, but **conservationists** argue that it should be left alone.

KEY

High mountains

Cold desert where the average temperature is less than 0 °C

Coniferous forest where the average temperatures are between 0 °C an[d]

▲ Active volcano

Major river

Major gorge

Large marshland

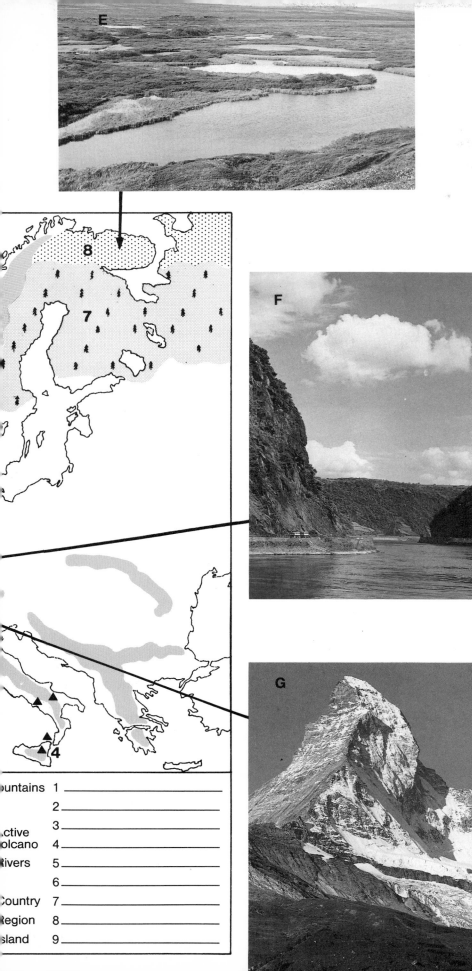

E

F

G

›untains	1 _____
	2 _____
ctive	3 _____
olcano	4 _____
Rivers	5 _____
	6 _____
Country	7 _____
Region	8 _____
sland	9 _____

1 Using a blank map of Europe, carefully make your own copy of map A and its key. Use an atlas to help you complete the key. (Hint: a physical map of Europe will be most useful.)

2 Look at map A and estimate what percentage of Europe's land surface is made up of wild landscapes. Is it 10%, 40%, 70% or 90%?

3 Each of photographs B–G shows a wild landscape in Europe. For each photograph, write a few sentences to explain why you think it is difficult for people to live there.

4 Which of the six wild landscapes shown would be easiest to change and live in? In a few sentences give reasons for your choice.

5 Copy the diagram below. Use an atlas to help you to write in the heights of Europe's highest mountains.

H

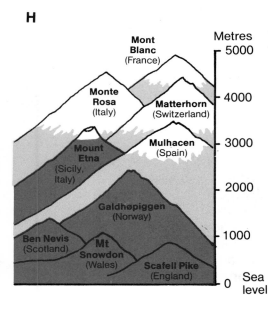

Mont Blanc (France)
Monte Rosa (Italy)
Matterhorn (Switzerland)
Mulhacen (Spain)
Mount Etna (Sicily, Italy)
Galdhøpiggen (Norway)
Ben Nevis (Scotland)
Mt Snowdon (Wales)
Scafell Pike (England)

Metres
5000
4000
3000
2000
1000
0 Sea level

A

B

C

The earth's natural landscapes and wild areas never seem to change – or do they? The fact is that things are always on the move. The only reason we don't notice is because the change is usually very slow.

There are three main kinds of change in the landscape. One is called **erosion**. This means the 'wearing down' of the landscape by such things as rivers, glaciers and the sea. Photo A shows that erosion can sometimes change things very quickly. But, most of the time, the erosion of the land happens very slowly. Glaciers take thousands of years to wear away and to widen the sides of valleys, for example.

Photo B shows another kind of change. Rocks, boulders and stones are being moved or 'transported' by the raging river. This movement is called **transportation**. In hot dry areas, rock particles (or sand) are often moved great distances by the wind when sandstorms take place.

Photo C shows a third kind of change called **deposition**. Sand has been dumped or deposited by the sea to form a sandspit. Sand can also be deposited by the wind to form dunes, or deposited by the sea to form beaches.

The wind, rivers, glaciers and the sea are all **agents of erosion**, transportation and deposition. They erode, transport and deposit rocks in the landscape. This is happening all the time and can be called a cycle (see picture E).

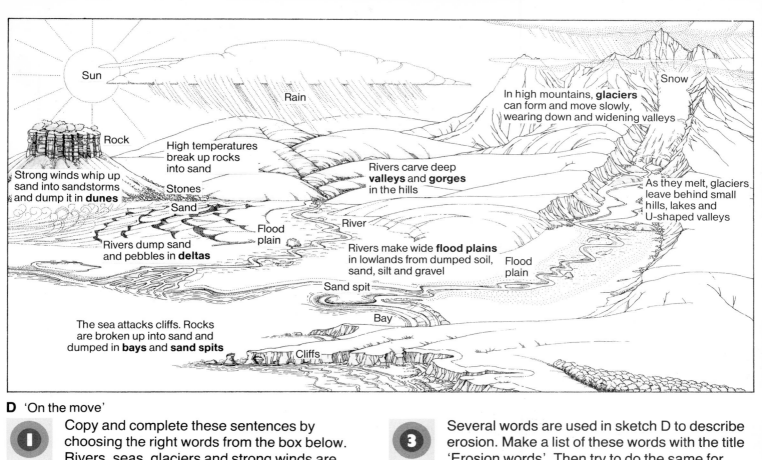

The labels on the landscape illustration read:

Sun

Rain

Rock

High temperatures break up rocks into sand

In high mountains, **glaciers** can form and move slowly, wearing down and widening valleys

Snow

Strong winds whip up sand into sandstorms and dump it in **dunes**

Stones

Sand

Rivers carve deep **valleys** and **gorges** in the hills

As they melt, glaciers leave behind small hills, lakes and U-shaped valleys

Rivers dump sand and pebbles in **deltas**

Flood plain

River

Rivers make wide **flood plains** in lowlands from dumped soil, sand, silt and gravel

Flood plain

Sand spit

The sea attacks cliffs. Rocks are broken up into sand and dumped in **bays** and **sand spits**

Bay

Cliffs

D 'On the move'

1 Copy and complete these sentences by choosing the right words from the box below.
Rivers, seas, glaciers and strong winds are
_____ __ _____. Each one has _____ main tasks.
These are erosion, _____ and _____
which change the landscape _____ _____.

> **Choose from:** three, very slowly, deposition, agents of erosion, transportation

2 Study photos A, B and C, then copy and fill in this table to explain what each photo shows.

3 Several words are used in sketch D to describe erosion. Make a list of these words with the title 'Erosion words'. Then try to do the same for deposition and for transportation.

4 Copy and complete diagram E by using the information shown in sketch D.

5 For photo A or B, write a newspaper story to describe what has happened. Choose either 'Cliffhanger!' or 'Boulders on the Move' as your headline and include at least one labelled sketch.

Photo	Agent of erosion	Main task

E

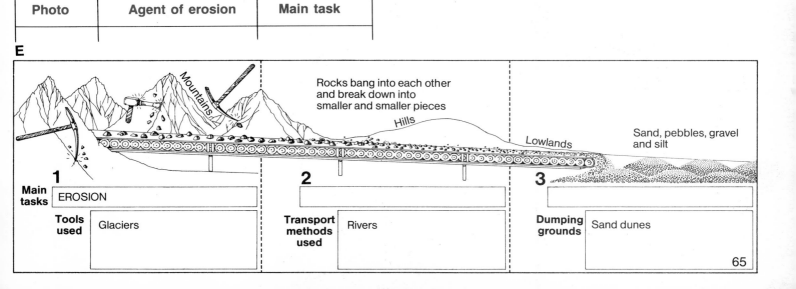

Mountains

Rocks bang into each other and break down into smaller and smaller pieces

Hills

Lowlands

Sand, pebbles, gravel and silt

	1	2	3
Main tasks	EROSION		
Tools used	Glaciers	**Transport methods used** Rivers	**Dumping grounds** Sand dunes

65

29 A NICE LANDSCAPE

A

During glaciation

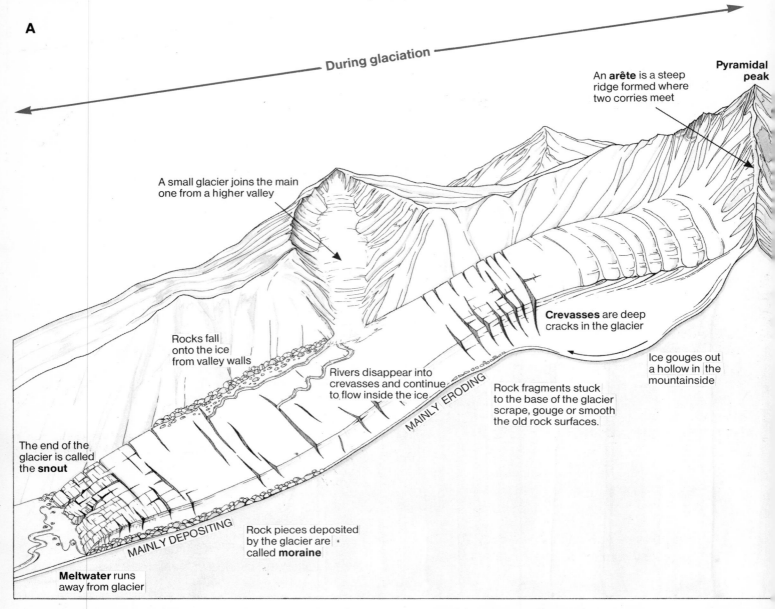

Pyramidal peak

An **arête** is a steep ridge formed where two corries meet

A small glacier joins the main one from a higher valley

Crevasses are deep cracks in the glacier

Rocks fall onto the ice from valley walls

Rivers disappear into crevasses and continue to flow inside the ice

Ice gouges out a hollow in the mountainside

MAINLY ERODING

Rock fragments stuck to the base of the glacier scrape, gouge or smooth the old rock surfaces.

The end of the glacier is called the **snout**

MAINLY DEPOSITING

Rock pieces deposited by the glacier are called **moraine**

Meltwater runs away from glacier

If you could look down on the earth from outer space, you would see that large parts of the land and sea are covered in ice or snow. In the very high mountains and in the north and south Polar regions the temperature stays low enough to stop the snow and ice from melting, even in summer. It is in these very cold areas that **glaciers** and **ice sheets** are found all year round. During the **ice ages** much more of the earth was covered in snow and ice.

An ice sheet is a glacier spread over a very large area. In most high mountain areas, glaciers are found only in the valleys and are called **valley glaciers**. When enough ice has built up, a glacier starts to flow very slowly. As it moves, it picks up small pieces of rock

which become frozen to the base of the ice. These scrape and grind away the land under the ice, rather like sandpaper wearing away wood. This **glacial erosion** can completely change a landscape, as shown in sketch A above. The eroded rocks, soil and boulders are **transported** (moved) under and in the ice. Other rocks and the **debris** from the valley sides are transported on top of the ice. When the ice melts, these rocks and debris are **deposited**. The deposit is called **moraine**.

Landscape features of glacial erosion and glacial deposition can be seen in many parts of the world. They tell us that an area has at some time been under the influence of ice.

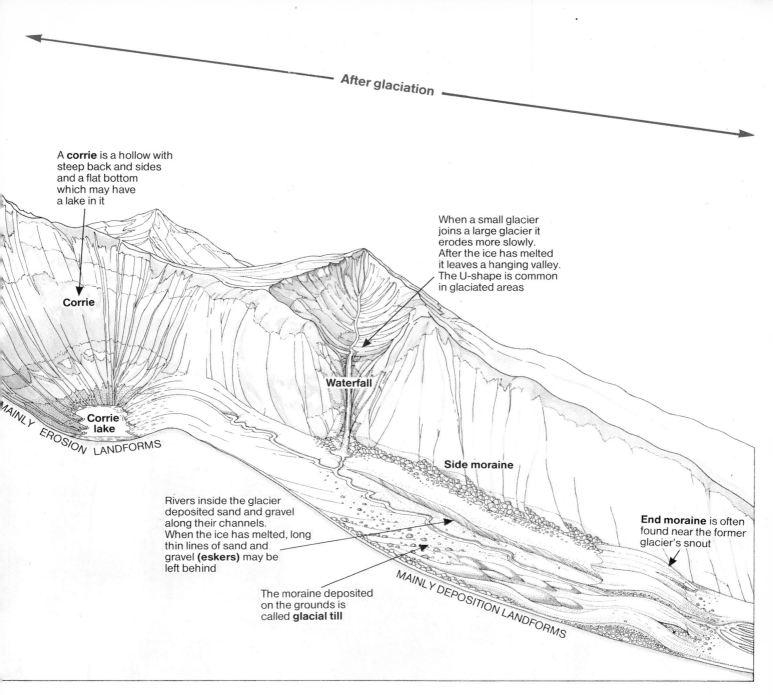

After glaciation

A **corrie** is a hollow with steep back and sides and a flat bottom which may have a lake in it

Corrie

When a small glacier joins a large glacier it erodes more slowly. After the ice has melted it leaves a hanging valley. The U-shape is common in glaciated areas

Waterfall

MAINLY EROSION LANDFORMS

Corrie lake

Rivers inside the glacier deposited sand and gravel along their channels. When the ice has melted, long thin lines of sand and gravel (**eskers**) may be left behind

The moraine deposited on the grounds is called **glacial till**

Side moraine

End moraine is often found near the former glacier's snout

MAINLY DEPOSITION LANDFORMS

 ❶ Study sketch A, then copy and complete the table below.

❷ Write a short description of each of the features you have listed in your table.

❸ Imagine that the ice age has returned to Europe. Write a report for a newspaper with the heading 'Europe in the grip of ice!'. Your report should describe the effects on farming, industry and people.

Features of glacial erosion	Features of glacial deposition
corrie	glacial till

30 A DISAPPEARING LANDSCAPE

Limestone is an unusual rock. Some very spectacular scenery is found in limestone areas, both above and below the ground.

There are two main types of limestone: hard and soft. It is hard limestone areas which have the most dramatic scenery (diagram A). This is because hard limestone dissolves easily in rainwater and in stream and river water. It is also because the rock has many **joints** (lines of weakness) in it. Water runs into these joints and dissolves the rock to form cracks and tunnels. These tunnels may run for several kilometres below the ground, and many of them open out into caves (diagram A). It is the excitement of finding caves that leads pot-holers into some very dangerous places!

Underground streams and rivers do not flow below the ground forever. When the layer of limestone comes to an end and another type of rock starts, the stream stops going underground and flows along between the two rock layers. It eventually reappears at the surface as a **spring**.

Chalk is a soft type of limestone. Chalk is not jointed like hard limestone but water can pass through it very easily, like a sponge. This means that, in chalk areas, there are very few streams on the surface and no underground tunnels or caves. When streams reach a layer of chalk, they disappear underground as the water sinks through the rock.

Very little soil is formed from hard limestone so limestone areas are not good for farming. Chalk makes a thin soil which can be used for growing grain crops and grass for grazing animals.

Limestone has been used for building for thousands of years. Nowadays, limestone is used in two ways. The stone can be used as building blocks and it can be crushed to make a powder which is used in making cement.

A An upland area of hard limestone

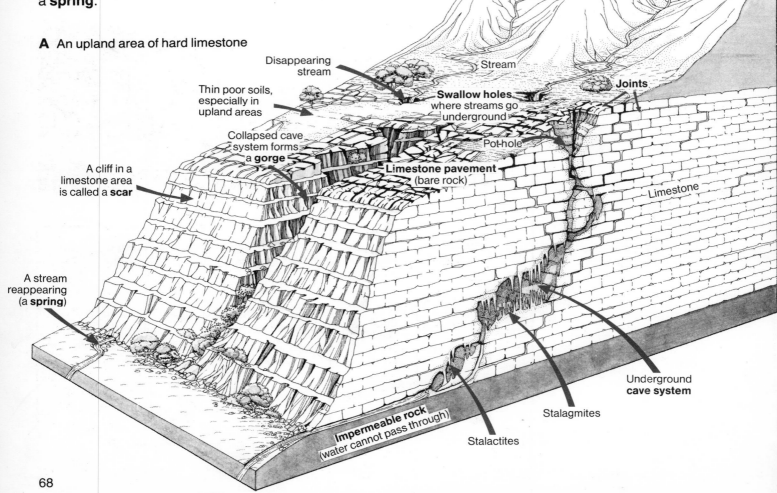

Impermeable rock (water cannot pass through)

Stream

Disappearing stream

Joints

Thin poor soils, especially in upland areas

Swallow holes where streams go underground

Pot-hole

Collapsed cave system forms a **gorge**

Limestone pavement (bare rock)

A cliff in a limestone area is called a **scar**

Limestone

A stream reappearing (a **spring**)

Impermeable rock (water cannot pass through)

Stalactites

Stalagmites

Underground **cave system**

Blue Circle is the name of the biggest cement-producing company in Europe. Five of its biggest works in Britain are shown on map B. The works are built next to the limestone quarries and most of them are in areas of beautiful scenery. For example, the Weardale and Hope Works are both in National Parks. The Blue Circle Company have to make sure that their works spoil the landscape as little as possible. To do this, they plant trees round their buildings and filter the dusty smoke from the chimneys. When the quarries are finished, they must be filled in and the soil put back so that grass and trees can be planted.

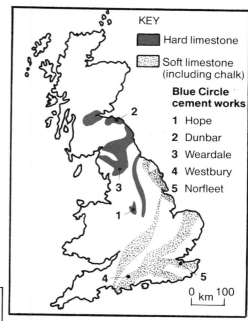

KEY

- Hard limestone
- Soft limestone (including chalk)

Blue Circle cement works
1 Hope
2 Dunbar
3 Weardale
4 Westbury
5 Norfleet

0 ___ km ___ 100

B Blue Circle cement works in Britain

 1 Copy and complete the table below, using diagram A to help you.

Some features of upland hard limestone scenery

Feature	Description
	caused when a cave system collapses
Swallow hole	
	natural cracks in the limestone
Underground cave	
	an area of bare limestone in an upland area

2 Write down two differences between chalk scenery and hard limestone scenery.

3 Copy and complete the table below using an atlas and information in map B to help you.

Name of cement works	Name of area where found	Rock type
1	Peak District	
2	East Lothian	
3	Yorkshire Dales	
4	Wiltshire	
5	Kent	

C Dunbar works

4 Photograph C shows the Blue Circle quarry at Dunbar. Describe what letters a–f show, using sketch D to help you.

D Quarrying limestone

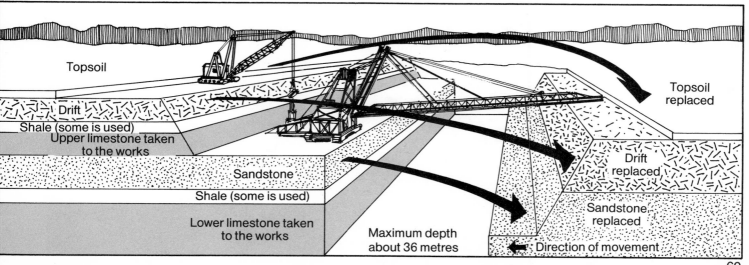

Topsoil

Drift

Shale (some is used)

Upper limestone taken to the works

Sandstone

Shale (some is used)

Lower limestone taken to the works

Maximum depth about 36 metres

Topsoil replaced

Drift replaced

Sandstone replaced

← Direction of movement

31 LAST LINE OF DEFENCE

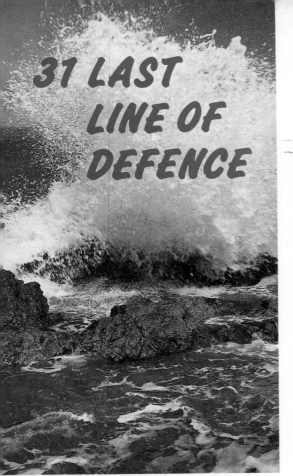

A Wave power

Coasts are fascinating places. Breakers crashing onto cliffs, surfers defying the waves, or holiday makers sunbathing on a sandy beach: all these can happen where land meets the sea.

Coasts are always changing. In some places, the coastlines are under attack from the sea. As photo A shows, waves have enormous power which is increased when small pieces of rock and sand are mixed up in the water. These wear away the rock of the coastline. Diagram B shows a coastline of **erosion**, where the sea is eating away land. Many of the familiar features which you can see at coasts are caused by this.

Eroded material may stay close to the cliff that it came from. But sometimes it is moved further up the coast (as shown in diagram C) and **deposited** by the sea to form new land. This is how beaches are formed.

An arch roof may collapse to leave a **stack**

At cracks or weaknesses in the rock, the waves may carve out **caves**

Steep **cliffs** are usually a sign of fast coastal erosion

Caves on either side of a **headland** (land jutting out into the sea) may join to form an **arch**

Waves, especially during storms, erode the bottom of the cliff. This **undercutting** makes the cliff unstable, and it collapses

A **wave-cut platform** is only seen at low tide. This is a sure sign that the sea is eating into the land. The rock platform was once the base of the cliff

B A coastline of erosion

The sea is not the only thing which wears away and erodes the coastline. People can also damage it. Coasts are attractive and popular spots to visit, and if too many people walk over the land they can wear down cliff paths, erode sand dunes, and damage steep bankings. It is expensive to protect these areas from this kind of damage (see photo E).

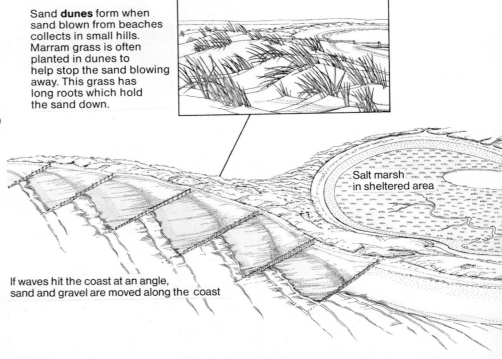

Sand **dunes** form when sand blown from beaches collects in small hills. Marram grass is often planted in dunes to help stop the sand blowing away. This grass has long roots which hold the sand down.

Salt marsh in sheltered area

If waves hit the coast at an angle, sand and gravel are moved along the coast

Lines of weakness in the rocks are picked out by the waves, making **coves** or **creeks**

D Needles, Dorset

E Resisting erosion by wind and feet

1 Make a list of the coastal features which you can see in photographs A, D and E. For each of the features in your list say whether it is a feature of erosion or deposition.

2 Copy and complete the table below, using diagrams B and C to help you.

Some features of coastal landscapes

Feature	Description
	a narrow rocky inlet
stack	
	a finger of land pointing into an estuary
wave-cut platform	
	a hollow carved out of seashore cliffs
sand dune	

3 Diagram F shows five coastal features, numbered 1–5. Write down the numbers 1–5 and beside each write the correct name of the feature shown.

coastline of deposition

River **estuary**

Sand moved by waves along the coast is often deposited in a **spit** when the sea meets quieter water in a river estuary. A spit is attached to the mainland

An offshore **sand bar** is separate from the mainland. It is deposited by the waves and can only be seen at low tide

F

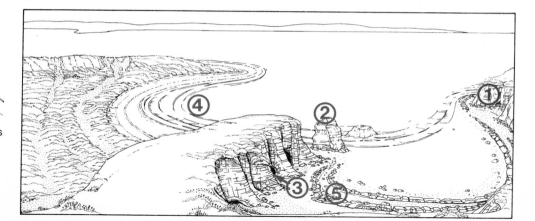

32 Crash, bang, wallop!

Most of the changes that take place in the natural landscape happen very slowly over a very long time. But sometimes, and usually without warning, change can take place very suddenly, such as a volcanic eruption or an **earthquake**.

These sudden and violent changes can cause massive damage and loss of life, as described in the newspaper stories in A. Such dramatic events are called **natural disasters**.

There are some parts of the world which are more likely to suffer from natural disasters than others. The earth's thin crust is made up of solid plates of rock which move around very very slowly. It is along the edges of these moving plates that most of the volcanic eruptions and earthquakes take place.

A

ARTHUR DAILY NEWS
BLACK CLOUD BLOTS SUN!
Ash found in Britain

12,000 dead in Agadir 'quake
Thousands buried in city of rubble

DAILY BLAH
Japan suffers landslides and floods

☒ DAILY DISC ☺
BOILING FURY!
Tourist town engulfed 112 die, flood drowns 32

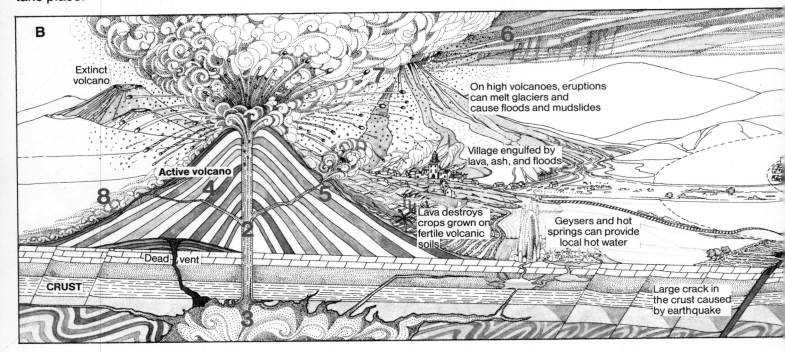

B

Extinct volcano

6

7

On high volcanoes, eruptions can melt glaciers and cause floods and mudslides

Village engulfed by lava, ash, and floods

Active volcano

8

4

5

Lava destroys crops grown on fertile volcanic soils

2

Geysers and hot springs can provide local hot water

Dead vent

CRUST

3

Large crack in the crust caused by earthquake

C The eruption of Vesuvius

Below the thin crust of the earth, the temperature is so high that some rock is molten. Where there are weak points in the solid rock nearer the surface, molten rock can force its way through cracks to form volcanoes (as shown in diagram B). The earth has over 500 **active** volcanoes and every year 30 or so erupt (photo C). Some volcanoes have never been seen to erupt. They are known as **extinct** (dead) volcanoes.

D Mexico City earthquake, 1985

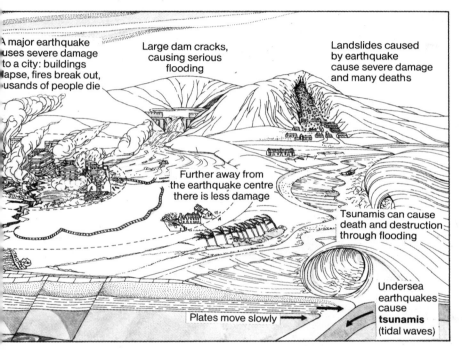

A major earthquake
causes severe damage
to a city: buildings
collapse, fires break out,
thousands of people die

Large dam cracks,
causing serious
flooding

Landslides caused
by earthquake
cause severe damage
and many deaths

Further away from
the earthquake centre
there is less damage

Tsunamis can cause
death and destruction
through flooding

Plates move slowly

Undersea
earthquakes
cause
tsunamis
(tidal waves)

An earthquake is a violent shaking of the ground, often caused by the build-up of great stresses in the earth's crust. Earthquakes usually happen along cracks (or faults) in the thin crust.

When the earth moves, buildings fall, cables and pipelines break, landslides occur and dams may burst to cause flooding.

Some earthquakes happen below the sea and may produce what are called **tsunamis**. Imagine facing a solid wall of water over 30 metres high and moving at a speed of over 500 km an hour! That is what a tsunami can be like.

Scientists study the movements along edges of plates to try to find out when eruptions or earthquakes are likely to happen. A warning may save many lives if people believe it and act on it. When the Nevado del Ruiz volcano erupted in Colombia in 1985, a warning was given but it was not taken seriously. The result was the loss of over 21 000 lives under a massive avalanche of mud, water and fallen buildings.

1 Write a sentence to explain what is meant by a 'natural disaster'.

2 The pictures on these two pages tell you about several kinds of natural disasters. Make a list of five of them.

3 In your own words explain the difference between an active and an extinct volcano.

4 The following labels are for the volcano diagram on picture B. Each label explains a different part of the eruption. Copy the labels and, beside each one, write the correct number from the volcano diagram.

Boiling **lava** pours down volcano
Magma rises through a pipe or **vent** in the volcano
Hot gases escape from a small **cone** on the side of the volcano
Lava, gas, dust, bombs and ash are erupted from the volcano's **crater**
Dust from the eruption blocks sunlight
Hot molten rock called magma fills giant chambers in the crust
Red-hot lava bombs thrown out of crater
Layers of ash and lava build up at each eruption

5 Look at picture B.
(a) Explain why volcanic eruptions sometimes lead to floods.
(b) Write down two good effects of volcanoes.
(c) In your own words, explain why earthquakes can lead to flooding.

33 Plough the fields and scatter

Farming is a primary industry. It produces food from the land.

In some ways, a farm is like an open-air factory. As shown in diagram A, a farm needs **inputs** to produce **outputs** which are then sold to markets and shops. Farming is a business and farmers run their farms to make a **profit** (see B). No two farms are exactly alike and each farmer must decide how to use the land to make the most profit.

One of the inputs shown in diagram A is 'E.E.C. farming policy'. This refers to the European Community of countries, including Britain. The Community gives special grants of money to European farmers to help and encourage them to produce the goods the Community needs.

There are millions of farmers in the European Community. They range in type from crofters in northern Scotland to olive growers in the Mediterranean lands. Altogether, Europe's farms produce enough food to feed the whole population *plus* an extra 300 million people. Much of this extra food is sold to other countries, but some of it is wasted.

A 'Open-air factories': farming

Inputs

Sunshine, warmth

Rain

Soil

Fertilisers

Seeds

Feed

Rent and buildings

Machinery

Workers

E.E.C. farming policy

TOMARKET LTD

Transport costs

Farm

Outputs

Milk

Vegetables

Meat

Grains

Profit

B Profit and loss

This farm sells produce to make money. It sells grain, animals and other products such as milk, eggs and cheese

MILK

MOOOOO

HONK HONK

Grains

The farm's costs are many. Fertiliser, seed, rent, power, new machinery and workers' wages must all be paid for

Rents

Power

Machinery

Seed

Feed

SALES — COSTS
minus

C A hill farm in northern England

Farming is always a risky business because it depends on things beyond the farmer's control: for example, the weather, plant and animal diseases, and the demands of consumers. Even so, some farmers have a tougher job than others.

Photo C shows a typical sheep farm in northern England. The farm covers many hectares of land, much of which is useless for growing crops. The weather is cold, wet and windy for several months of the year; slopes are steep and soils are poor. Only the flatter better drained soils on the low ground can be used for growing fodder crops which are fed to the sheep in winter. Areas like this could not be farmed unless the farmers were given extra help by the government (see Table 1).

1 Farming is a primary industry. Write a sentence to explain what is meant by the term 'primary industry'.

2 Using diagram A, make a list of the outputs and inputs of farming.

3 The hill farmer grows turnips and oats as fodder crops. What are fodder crops used for?

4 Explain what is meant by profit. Use B to help you.

5 The hill farmer keeps 200 sheep (for wool) and 10 dairy cattle (milk is sold). Make a list of the inputs and outputs of a hill farm.

6 What do you think should be done with the extra food Europe produces? Give reasons for your answer.

7 Copy out and complete Table 1 for a hill farm. What makes it possible for the hill farmer to make a profit?

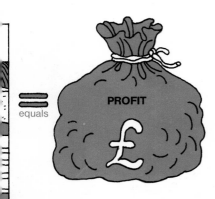

equals   PROFIT £

Table 1 Costs and sales for a hill farm

Costs		Returns from sales	
Seed	£1000	Wool	£4500
Fodder	£1000	Milk	£2000
Fertiliser	£ 500	Government subsidy	£4500
Electricity and water	£1000		
Repairs	£ 500	Total sales:	£
Wages	£5000		
		Subtract	
Total costs:	£	Total costs	£
		Profit:	£

34 Sold down the river

Map A shows the whole length of the River Tees in north-east England, from its source in the Pennine Hills to its mouth on the North Sea coast. The River Tees, like most rivers, can be divided up into three parts or courses. In the **upper course** in the hills, the main task of the river is **erosion** (wearing down of rock). This can often result in spectacular scenery such as gorges (sketch B) and waterfalls (photo D). Up in the hills, the weather is usually cool, windy and wet and the valley sides are steep and rugged. It is very difficult to grow crops, but sheep farming is common, and so is forestry.

At one point in the upper course of the Tees Valley, a dam has been built to form the Cow Green Reservoir (photo C). The reservoir water is carried by pipeline to be used in the farms, villages and towns in the lowland part of the valley.

In the **middle course** of the (sketch E), the valley becomes wider and the land lower. The Tees flows more slowly and big bends (or **meanders**) are formed. Sheep and beef cattle graze on the pastureland of the valley floor.

As the river gets closer to the sea, it flows much more slowly across the flat **flood plain**. This is the **lower course** of the river (sketch H). The weather is warmer here, and the land lower, flatter and more fertile. There are many farms, villages and market towns (such as Yarm in photo G) and the landscape is covered in fields where dairy cattle graze and grain crops and vegetables are grown.

As the Tees gets even closer to the sea, industrial towns such as Middlesbrough and Stockton take over from the farmland. Finally, at the river mouth, there is a huge chemical works and other industries which cover large areas of land (sketch F).

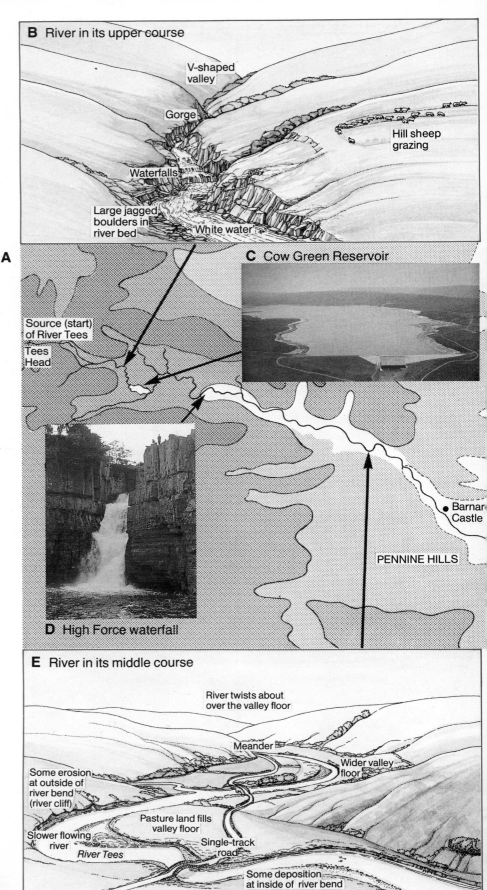

B River in its upper course

V-shaped valley

Gorge

Hill sheep grazing

Waterfalls

Large jagged boulders in river bed

'White water'

A

C Cow Green Reservoir

Source (start) of River Tees

Tees Head

Barnard Castle

PENNINE HILLS

D High Force waterfall

E River in its middle course

River twists about over the valley floor

Meander

Wider valley floor

Some erosion at outside of river bend (river cliff)

Pasture land fills valley floor

Slower flowing river

River Tees

Single-track road

Some deposition at inside of river bend

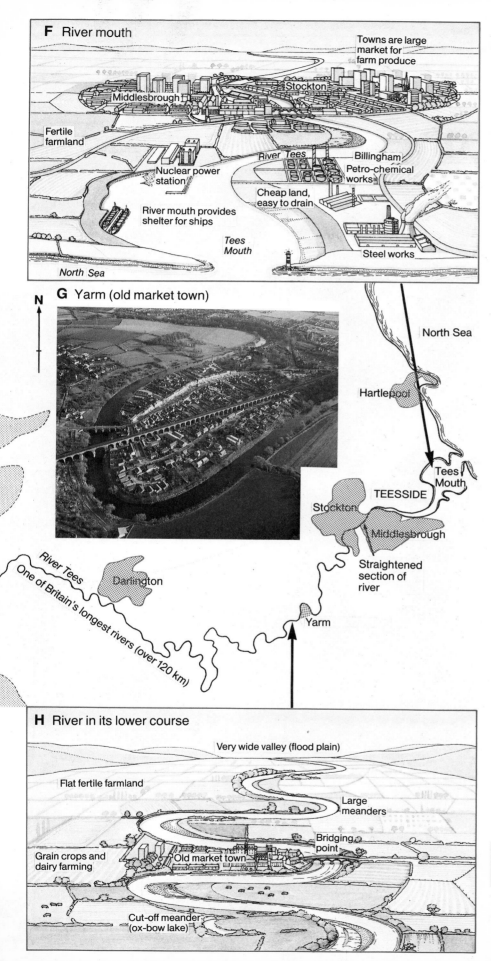

F River mouth

Towns are large market for farm produce

Middlesbrough

Stockton

Fertile farmland

Nuclear power station

River *Tees*

Billingham Petro-chemical works

Cheap land, easy to drain

River mouth provides shelter for ships

Tees Mouth

Steel works

North Sea

G Yarm (old market town)

N

North Sea

Hartlepool

Tees Mouth

TEESSIDE

Stockton

Middlesbrough

Straightened section of river

River Tees
One of Britain's longest rivers (over 120 km)

Darlington

Yarm

H River in its lower course

Very wide valley (flood plain)

Flat fertile farmland

Large meanders

Bridging point

Grain crops and dairy farming

Old market town

Cut-off meander (ox-bow lake)

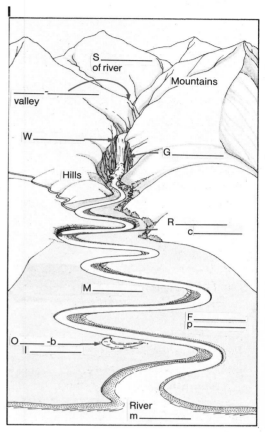

1 Copy and complete the sketch below of a river valley by naming the river features. (Hint: see sketches B, E, F and H.)

I

S ____ of river

Mountains

____ valley

W ____

G ____

Hills

R ____
c ____

M ____

F ____
p ____

O ____ -b ____

l

River m ____

2 Copy and complete the table below by studying the pictures on these two pages. (Land use means what people do with the land.)

Part of river	Land use
Upper course	
Middle course	
Lower course	
River mouth	

3 Write a paragraph to describe what changes in land use you would see as you travel down a river valley from source to mouth.

35 Gruyère to Gouda

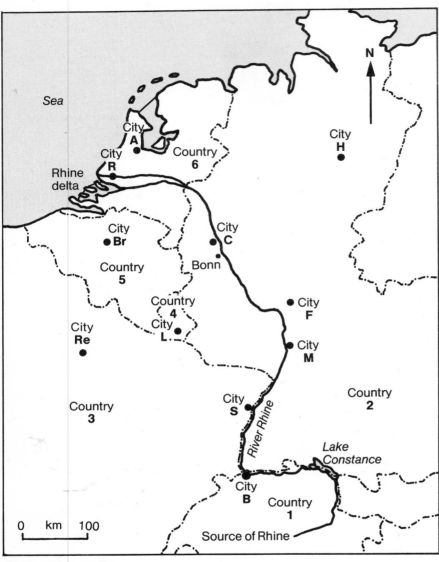

A The Rhine valley

Map labels:
- Sea
- City A
- City R
- Rhine delta
- City H
- Country 6
- City Br
- City C
- Bonn
- Country 5
- Country 4
- City L
- City Re
- City F
- City M
- Country 3
- City S
- River Rhine
- Country 2
- Lake Constance
- City B
- Country 1
- Source of Rhine
- N
- 0 km 100

B Alpine dairy farming

If you could travel all the way from the source to the mouth of the River Rhine, you would pass through four countries and six large cities (see map A).

The total length of your journey would be over 1325 km from the wild mountain landscape of the Swiss Alps to the artificial landscape of the Dutch **polders**.

On its long journey, the Rhine shapes the natural landscape. It has eroded (worn down) steep valleys and gorges in mountains, and deposited silt to form broad fertile plains in the lowlands, and a wide delta at its mouth.

Although these landscapes are very different, they do have some things in common – and one of them is cheese! Swiss, French and Dutch cheeses are very famous all over the world and some of them are made in different parts of the Rhine valley. In the Alps of Switzerland (photo B), dairy cows graze the mountain pasture in summer and their milk is used to make Gruyère and Emmental cheeses. At the other end of the Rhine valley, dairy cows graze on the rich fertile lowlands of the Rhine delta in the Netherlands (photo C). Here, their milk is made into Gouda and Edam.

The cheeses may be similar, but not the landscape they come from!

C Polder dairy farming

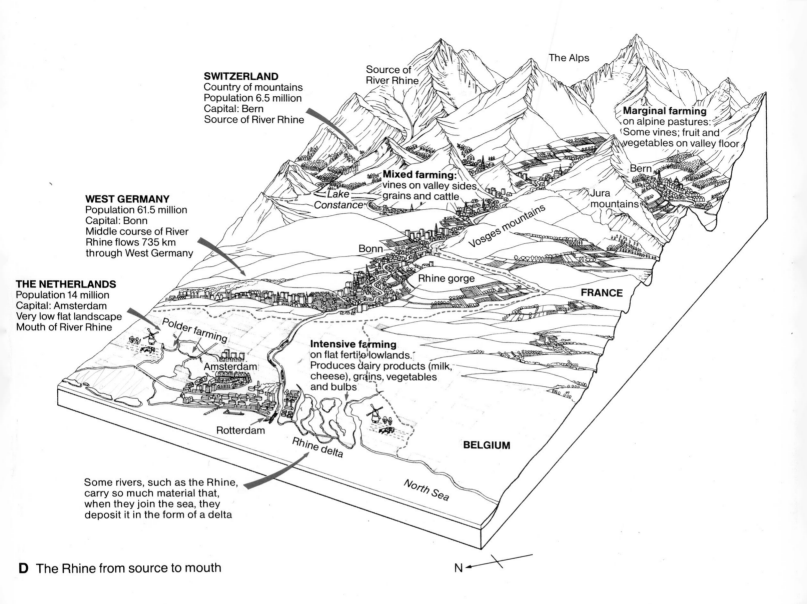

SWITZERLAND
Country of mountains
Population 6.5 million
Capital: Bern
Source of River Rhine

Source of
River Rhine

The Alps

Marginal farming
on alpine pastures.
Some vines; fruit and
vegetables on valley floor

Bern

WEST GERMANY
Population 61.5 million
Capital: Bonn
Middle course of River
Rhine flows 735 km
through West Germany

Mixed farming:
vines on valley sides,
grains and cattle

Lake
Constance

Jura
mountains

Vosges mountains

Bonn

Rhine gorge

FRANCE

THE NETHERLANDS
Population 14 million
Capital: Amsterdam
Very low flat landscape
Mouth of River Rhine

Polder farming

Amsterdam

Intensive farming
on flat fertile lowlands.
Produces dairy products (milk,
cheese), grains, vegetables
and bulbs

Rotterdam

Rhine delta

BELGIUM

Some rivers, such as the Rhine,
carry so much material that,
when they join the sea, they
deposit it in the form of a delta

North Sea

D The Rhine from source to mouth

N

Use an atlas to help you complete a copy of
map A. (Hint: Make a key below your map for
the names of the countries, cities and sea area
on the map.)

Using the scale bar on map A, work out how
many kilometres it is between
(a) the source of the Rhine and its mouth at
Rhine delta;
(b) Basle and Rotterdam;
(c) Reims and Hanover.

Briefly describe the different landscapes linked
by the River Rhine in photos B and C.

Copy and complete the table below by using
diagram D.

Stage of River Rhine	Country	Scenery	Land use
Upper course			Alpine farms
Middle course		Rhine gorge	
Lower course			
Mouth of river	Netherlands		

79

36 Life at the top

FRANCE

WEST GERMANY **Schaffhausen**

Rhine

Basle

A

Basle: 270 metres

Average monthly temperature (°C)

20
10
0
−10

J F M A M J J A S O N D

Yearly number of days with frost: 67
Yearly number of days with snowfall: 19

The map on this page shows the **upper course** of the River Rhine. For most of its length from high in the Alps to the Swiss city of Basle, the Rhine marks the frontier between Switzerland and Liechtenstein, Austria and West Germany.

For much of its journey through the Swiss Alps, the Rhine flows in a deep glaciated valley (photo B). Valleys like this are difficult to use for farming. The valley sides are too steep for crop growing, and they block the sunshine from parts of the valley floor. Crops need sunshine and warmth, so very few are grown on the shady side of the valley.

B A typical Swiss valley

△ 3620

△ 3328

Sedrun

△ 32

N

C

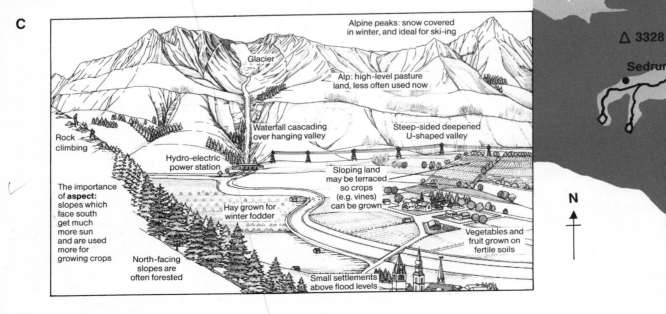

Alpine peaks: snow covered in winter, and ideal for ski-ing

Glacier

Alp: high-level pasture land, less often used now

Waterfall cascading over hanging valley

Steep-sided deepened U-shaped valley

Rock climbing

Hydro-electric power station

Sloping land may be terraced so crops (e.g. vines) can be grown

The importance of **aspect**: slopes which face south get much more sun and are used more for growing crops

Hay grown for winter fodder

Vegetables and fruit grown on fertile soils

North-facing slopes are often forested

Small settlements above flood levels

Swiss Alpine farmers have had to make best use of the little land they have. Sketch C shows how they have done this. Dairy farming is one of the most important types of farming in the upper Rhine valley. The cows need plenty of hay as food for the long hard winters, so much of the land is used for growing grass to make the hay. Lower down the valley, the main crops are fruit, vegetables and grapevines.

Farmers also make some extra money out of the tourist industry. They often provide accommodation for tourists, and even take winter jobs in the ski centres.

After the River Rhine leaves the Bodensee, it flows westwards to Basle. The land is lower, the valley wider and the climate less harsh here. Dairy and arable farming are still important but the main crops are wheat and potatoes.

1 Use the map scale to work out the length of the upper course of the Rhine (from Davos to Basle).

2 Use climate graphs A and D to help you explain why farming is easier around Basle than around Davos.

3 Use sketch C to describe how the Swiss have made use of the deep Alpine valleys.

4 Nestlés, Lindt and Tobler are the names of three Swiss food companies. Find out what they make, and explain how they depend upon the Alpine valley farms to make their products.

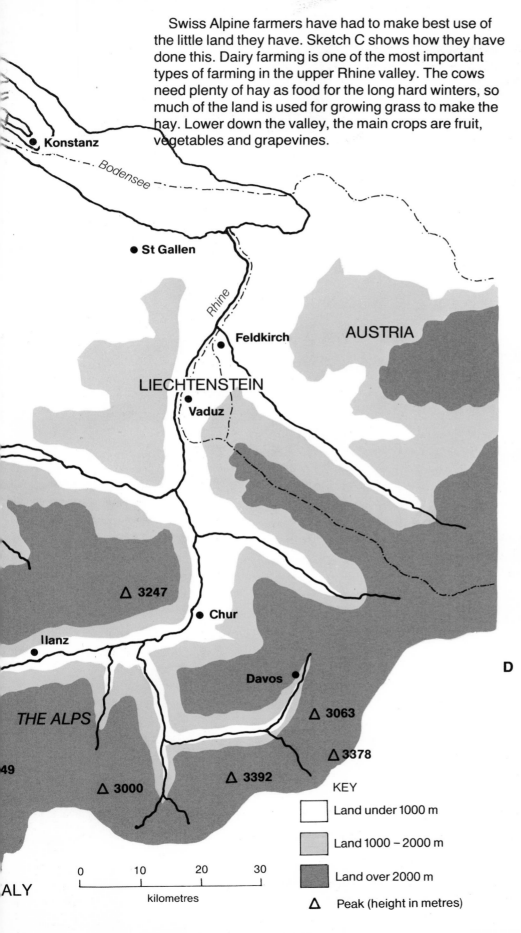

Konstanz

Bodensee

St Gallen

Rhine

Feldkirch

AUSTRIA

LIECHTENSTEIN

Vaduz

△ 3247

Chur

Ilanz

Davos

THE ALPS

△ 3063

△ 3378

49

△ 3000

△ 3392

ITALY

KEY

Land under 1000 m

Land 1000 – 2000 m

Land over 2000 m

△ Peak (height in metres)

0   10   20   30
kilometres

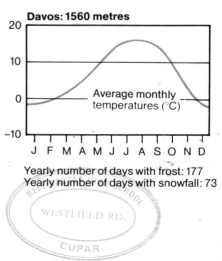

D

Davos: 1560 metres

Average monthly temperatures (°C)

J F M A M J J A S O N D

Yearly number of days with frost: 177
Yearly number of days with snowfall: 73

37 Down the Rhine

The **middle course** of the River Rhine is the stretch between the cities of Basle and Arnhem (map A). The river leaves Switzerland to form the border between France and West Germany, then it continues through West Germany until it crosses the Netherlands border just before Arnhem.

Between Duisburg and Bonn, the Rhine valley is mostly built up but, further south, there are fewer towns and industries, and the landscape is mainly rural.

Map A names the main crops grown and farming types in each area. As you can see, vines are common in the south and central areas. It is here that some of the world's most famous wines come from, such us Leibfraumilch, Mosel and Alsace.

Most of the vines are grown on terraces cut into the sloping valley sides (photo B). The slopes which face south or east are the best because they receive more sunshine and warmth than the north- or west-facing slopes.

Dairy and arable farming are more important than vineyards in the wider flatter parts of the valley (for example, near Mannheim). These areas produce plenty of milk products and other foodstuffs for the millions of people living in the northern industrial areas and the nearby cities.

The Rhine valley changes its shape several times between Basle and Arnhem. One of the most spectacular stretches is between Mainz and Koblenz where it flows at the bottom of a steep-sided valley called a **gorge** (photo E).

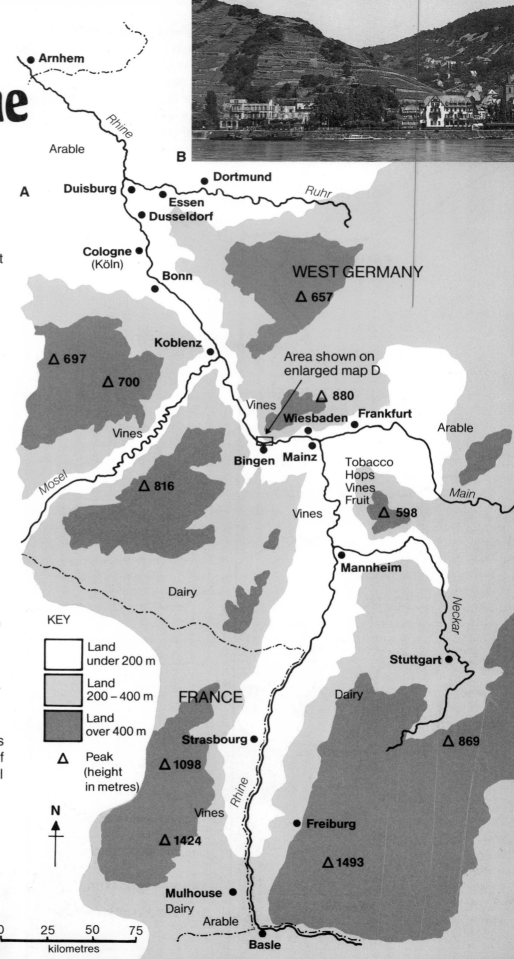

KEY

Land under 200 m

Land 200 – 400 m

Land over 400 m

△ Peak (height in metres)

N

0   25   50   75

kilometres

C

RHEINGAU

Schloss Schönborn

1979er

Rüdesheimer Berg Rottland

Riesling

Auslese

A. P. Nr. 31 052 035 80

Domänenrat

Erzeugerabfüllung Domänenweingut Schloss Schönborn **Hattenheim**
WEST-GERMANY

QUALITÄTSWEIN MIT PRÄDIKAT   750 ml

ETIENNE 1208

What the label says

1 The name of the region. Rheingau region is said to produce the best German wine. The vines are grown on the south-facing slopes of the Rhine between Mainz and Bingen (see map A).

2 When the wine was bottled: 1979 (vintage).

3 The name of the wine. This tells us the wine is from Berg Rottland vineyard near the town of Rüdesheim (see map D), is made from the Riesling grape, and is 'Auslese', which means the grapes are picked by hand and carefully selected.

4 The name of the producer.

5 *Qualitätswein mit Prädikat* means the wine is of the top quality.

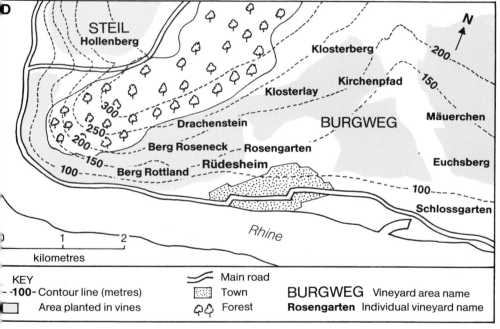

D

STEIL
Hollenberg
Klosterberg
Kirchenpfad
Klosterlay
Drachenstein
BURGWEG
Mäuerchen
Berg Roseneck — Rosengarten
Rüdesheim
Euchsberg
Berg Rottland
Schlossgarten
Rhine

N

200
150
300
250
200
150
100
100

0   1   2
kilometres

KEY
-100- Contour line (metres)
☐ Area planted in vines
〰 Main road
▦ Town
🌳 Forest
BURGWEG Vineyard area name
Rosengarten Individual vineyard name

E The Rhine Gorge

---

① Use the scale on map A to work out the length of the middle course of the Rhine (from Basle to Arnhem).

② Look carefully at photo B. What have farmers done to make it possible to grow vines on steep slopes?

③ Use maps A and D to describe the location of the vineyards near Rüdesheim (mention the river, towns and slopes).

④ Look at map D. What evidence is there to show that Rüdesheim is in a gorge? (Use photo E to help you).

⑤ Imagine you are in a boat on the Rhine as it passes the town of Rüdesheim. Try to describe the view you would see of the town and surrounding area. Use the map key to help you.

38 TURNING THE TIDE

As the River Rhine gets near to the North Sea, it begins to wander over a flat **flood plain**. This is its **lower course** (map A).

Before the river reaches the sea, it separates into several smaller branches. This very flat area is called a **delta**. It is made up of mud and fine soil which has been carried by the river from its upper and middle courses.

Most of the land in the Netherlands lies below sea level, but it has been reclaimed from the sea. Over many years, the Dutch have worked to drain the land and built walls to stop the sea from drowning it again. These drained areas are called **polders** (photo B and sketch C). The polder land is used for farming as well as building towns and industries.

B Dutch polders: reclaimed land

A

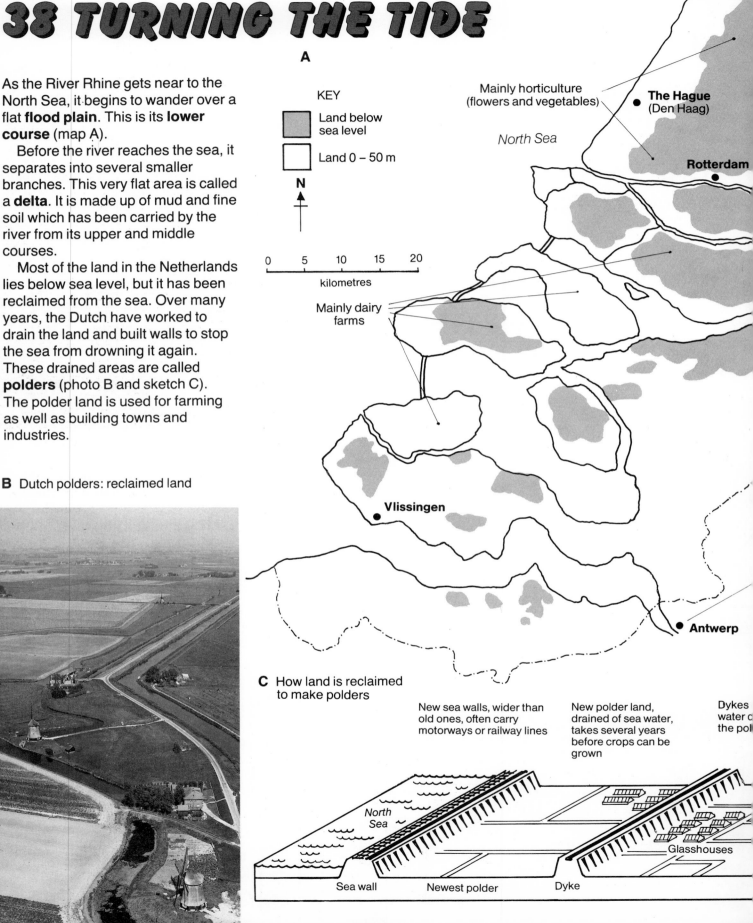

KEY

Land below sea level

Land 0 – 50 m

N

| 0 | 5 | 10 | 15 | 20 |

kilometres

North Sea

Mainly horticulture (flowers and vegetables)

The Hague (Den Haag)

Rotterdam

Mainly dairy farms

Vlissingen

Antwerp

C How land is reclaimed to make polders

New sea walls, wider than old ones, often carry motorways or railway lines

New polder land, drained of sea water, takes several years before crops can be grown

Dykes
water c
the pol

North Sea

Glasshouses

Sea wall   Newest polder   Dyke

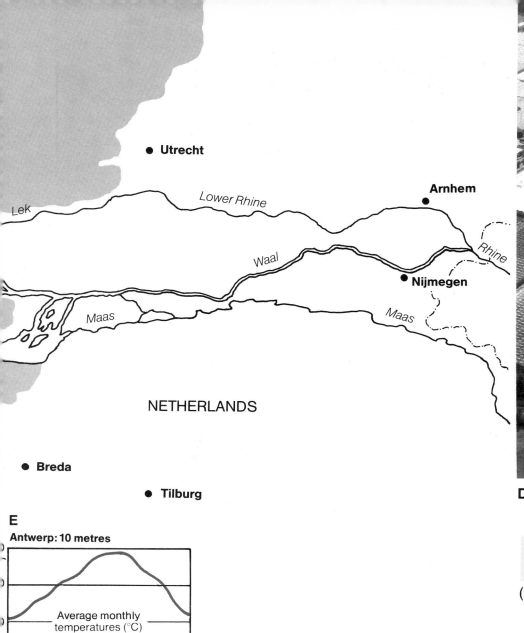

● **Utrecht**

Lek

Lower Rhine

Arnhem

Waal

Rhine

Maas

● **Nijmegen**

Maas

NETHERLANDS

● **Breda**

● **Tilburg**

E

Antwerp: 10 metres

Average monthly
temperatures (°C)

J F M A M J J A S O N D

Drainage ditches
between fields collect
water which is pumped
up to canals on the
dykes

The oldest polders are
usually smaller in
size than newer ones.
Water used to be pumped
by windmills but diesel
pumps are now used

Land above
sea level

Oldest polder

D Glasshouses, a common sight in
the Netherlands

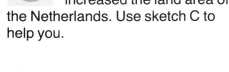

1 Use the map scale to work
out the length of the lower
course of the River Rhine
(from Arnhem to the North Sea).

2 Write a paragraph to
describe how people have
increased the land area of
the Netherlands. Use sketch C to
help you.

3 Name the features shown in
sketch C that are also
shown in photo A.

4 Photo D shows a typical
view in the Dutch polder
lands. Why do you think
glasshouses are so common in an
area which produces a lot of fruit and
vegetables?

85

39 Banks and Braes

Look at the Ordnance Survey map extract of Glenmore and the Cairngorm mountains on page 60.

The Cairngorm mountains in Scotland are the highest **mountain range** in Britain and one of the largest areas of **wild landscape** (photo A). Cairn Gorm mountain is one of the highest in the range. It rises to over 1245 metres (see grid square 0004 on the map extract).

The mountains are made of very old, high granite rocks. They have been eroded (worn down) by great ice sheets that once covered the whole area. You can see some of the features made by the ice if you look at the map extract. Look at the rounded mountain tops at grid squares 0005 and 9304. Now look for the large deep corries such as Coire an t-Sneachda (9903). These have been scooped out by ice and have very rugged rock walls. Ridges such as the Fiacall (9904) stand between the corries. Below the corries, valleys have U-shaped sides, made by moving glacier ice. These valleys are sometimes filled with long narrow (or ribbon) lakes, such as Loch Avon (0102).

A Cairn Lochan from Cairn Gorm summit

B Loch Morlich and the Cairngorm mountains from grid reference 9710

1 The Glenmore and Cairngorm area has a lot to offer visitors. Look carefully at the map extract (and photos B and C). Make a list of the different leisure activities shown.

2 In which part of the map do most leisure activities take place? Why?

3 Which two map features suggest that the Cairngorm mountains might be dangerous for hill-walkers?

4 Give a reason why the road in grid square 9807 may be closed to skiers in winter.

5 Write a few sentences to describe the landscape and likely leisure activities in each of the scenes shown in photos A, B and C.

6 (a) What tells you that the Glenmore Forest Park has been planted and is not a natural forest?
(b) Which type of tree has been planted in the Queen's Forest (see grid square 9610)?

7 Which grid squares have the following?
(a) The lowest land.
(b) The highest land.
(c) The deepest loch.
(d) The largest area of water.
(e) The most settlement.
(f) An unusual animal's home!

C Coire Cas from the chairlift

As the ice sheets melted they left behind many small hills of moraine (9508) in lowland areas. Since then, many rivers, such as the Beanaidh Bheag (9402) have continued to erode the Cairngorms. Today, Glenmore Forest Park (9710) is famous for **leisure activities**. Loch Morlich, with its Sandy Beach (photo B) caters for water sports. Over 5000 skiers use the ski lifts (photo C) in Coire Cas (9905) each day in the snow season. The mountains are also popular with walkers and climbers throughout the year.

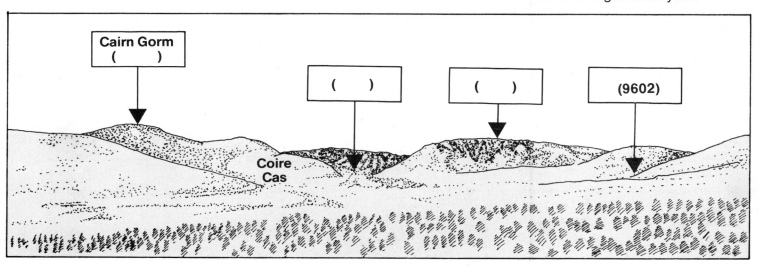

Cairn Gorm ( )

( )

( )

(9602)

Coire Cas

D Cairngorm mountains from Glenmore Youth Hostel (9709)

 8 Copy and complete the boxes in diagram D using the information given below. (Names and grid references are not correctly matched in the table.)

Names	Grid references
Cairn Lochan	(0004)
Lairig Ghru	(9903)
Cairn Gorm	(9602)
Coire an t-Sneachda	(9802)

 9 Study both the map extract and photos A, B and C. For each photo, work out the direction in which the photographer was looking.

10 Copy and complete Table 1 by using the information on the Glenmore map extract. (Hint: give four figure grid references and fill in the missing sketches, maps and names.)

Map feature	Name	Sketch	Map	Four-figure reference
Loch	Loch Morlich			
Steep-sided valley	Lairig Ghru			
Corrie	Coirean t-Sneachda			
Conical hill				9606
Ridge	Fiacaill a Choire Chais			

40 TIME OUT!

Everyone looks forward to their **leisure** time, especially going on holiday and becoming a tourist. People talk of wanting different things from their leisure time, such as: rest, fun, escape, action, 'paradise', interesting places, a sun tan or beautiful scenery.

People put the landscape to many different uses for their 'time out'. This can mean visiting theatres and discos at a resort (photo A); visiting ancient ruins (photo B); having fun on the beach (photo C); climbing mountains or bird-watching (photo D); skiing in winter (photo E).

'Time out' activities such as these can be seen as either **active** (for example, skiing) or **passive** (for example, sunbathing). Most people usually like a mixture of both in their leisure time.

Whatever their choice of leisure activity, most people agree that their 'time out' is very important for their health and well-being.

When tourists visit holiday areas, this can mean big business for hotels, shops and cinemas. But tourism can also cause problems such as traffic jams, noise, litter and damage to the landscape.

A Sunbathers, Weymouth (Dorset, England)

B The Acropolis, Athens (Greece)

C Surfing, Newquay

D The Scottish Highlands

E Skiing in the Swiss Alps

1 Tourism is an example of a service industry. Explain what this means.

2 Study photos A–E. What does each place offer the holiday maker? Write a short description for each photo.

3 Using a blank map of Europe and an atlas, mark on the places shown in photos A–E. Use the symbols below to make a key.

⊠	Historical centres
⛰	Mountains or winter resorts
▤	Coastal resorts (long stay)
●	Coastal resorts (day trips)

4 Complete a larger copy of the table below by listing the different types of leisure activities.

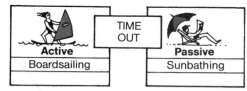

Active	TIME OUT	Passive
Boardsailing		Sunbathing

5 Why can the tourist industry be both a benefit and a headache for holiday places?

6 Look carefully at the information on these pages, then copy and complete the diagram and chart below (F).

F 'Time out' landscapes

Clear blue seas — Golden beaches — Holiday resort — Historical buildings — Old settlements — Home area — Countryside — Wildlife — Forests — High mountains — Volcano

Facilities	Hotels, swimming pools			Golf courses, caravan sites	
Activities (summer)				Fishing	
Activities (winter)			Cinema	Curling	
Example in Europe		Acropolis (Athens)			Cairngorm mountains

41 High days and holidays

Over four million people are employed in the **tourist industry** in Europe. That sounds a lot, but think about all the jobs that need to be done to cater for our holidays. Everything from selling travel tickets to cleaning hotel bedrooms.

200 years ago, holidays were almost unknown. Working people had almost no days off work and not enough money to go away. Only the wealthy could afford holidays. They often went to spa towns to 'take the waters'. Leamington and Bath, in England, are both spa towns. They used to be very popular holiday resorts. People travelled to them in horse-drawn carriages and stayed in boarding houses or rented rooms.

By the 1850s (just over 100 years ago), more people were taking holidays. Railways had been built and seaside resorts grew up around Britain's coast. Southport, Brighton and Eastbourne were all popular resorts for the wealthy in those days.

Early this century, most people took some sort of holiday away from home at least once a year. Even so, most working people only had Sundays off. In the summer, they often took day trips by bus or train to the nearest seaside resort. Scenes like the one in photo A were common in the 1920s and 1930s.

A Bank holiday crowds, Margate, 1935

B

Since those early days, there have been many big changes. Jet planes, motorways, longer holidays with pay, and higher wages are just some of them. All these changes have made it possible for people to travel farther away from home for their holidays.

Nowadays the most popular holiday is the long-stay 'sun, sea and sand' holiday (photo C). Then there are winter sports holidays in the mountains.

Map D shows where most British people take their holidays. Spain is the most popular choice of country for British tourists, and tourism is Spain's largest industry.

C Benidorm: Europe's most popular long-stay resort

D Where the British take their holidays

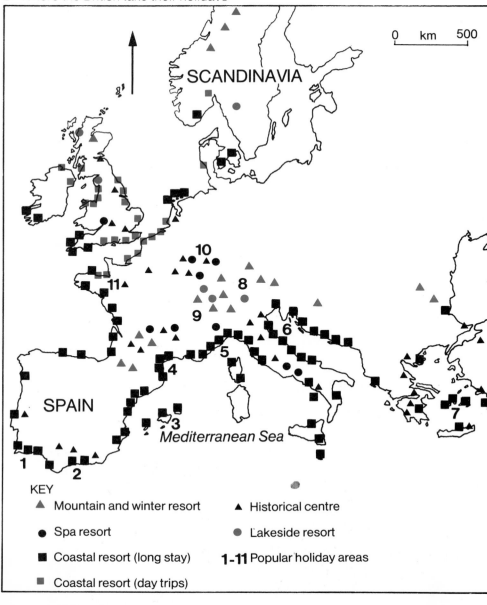

0 km 500

SCANDINAVIA

SPAIN

Mediterranean Sea

10
11
8
9
6
5
4
3
7
1
2

KEY

▲ Mountain and winter resort   ▲ Historical centre

● Spa resort   ● Lakeside resort

■ Coastal resort (long stay)   **1-11** Popular holiday areas

■ Coastal resort (day trips)

1 The tourist industry employs millions of people. Make a lists of at least ten jobs that are part of the tourist industry.

2 What types of holiday are most popular today? Why? (Hint: use diagram B to help you.)

3 Study photo C. What can you see in the photo that tells you Benidorm is a holiday resort? Make a short list.

4 Write a report called 'The growth of holidays'. (Hint: mention dates, holiday areas, transport, holiday makers, etc.)

5 Study map D then make an extra key to name holiday areas 1–11 shown on the map. Use an atlas to help you to find the names of the 11 areas. Choose from the names in the box below.

Choose from:
Majorca, Alps, Algarve, Loire Valley,
Costa Brava (Barcelona),
Italian Riviera, Greek Islands,
Rhine Valley, Costa del sol (Malaga),
Tyrol (Austria), Ligurian Riviera

6 Look carefully at map D.
(a) Where are most mountain or winter holiday resorts? Why?
(b) Where are most long-stay resorts? Why?
(c) Which parts of Europe do British people visit for day trips or short-stay holidays?
(d) Scandinavia is not very popular with British holiday makers. Why might this be so?

Investigating

Carry out a survey within your class. Find out where people have been on holiday over the past three years and for how long their holidays lasted.

42 Weather here, wish you were lovely!

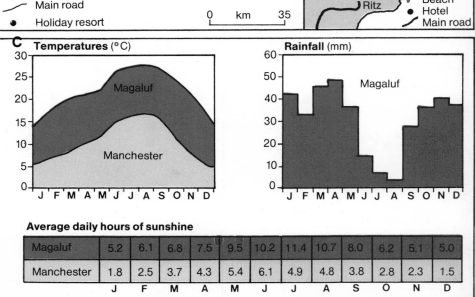

A

For lively family fun, Magaluf is hard to beat (photo A). By day and by night there is plenty to amuse, delight and entertain.

This popular Spanish resort is located on the largest of the Balearic Islands, Majorca (map B) and lies just 16 km from the capital, Palma.

The island basks in hot dry sunny summers (diagram C). It is famous for its beautiful beaches, castles, markets, shops and olive groves. The hotels, Magaluf Ritz and Flamenco Sol, are situated beside 1 km of glorious sandy beach in the sweeping bay of Palma (map B).

There is an excellent bus service to Palma and all parts of the island. Eating out can vary from snack bars to luxurious restaurants. Bars, cafés and nightspots abound. Sports enthusiasts can play golf nearby; play tennis or go horse riding in the resort itself, and enjoy a wide variety of water sports.

Descriptions of holidays such as this can be found in many travel brochures. Each year, the Mediterranean coast of Europe is visited by over 100 million holiday makers.

The 'sun, sea and sand' holiday is the most popular type of summer break.

The Spanish resort areas such as the Costa Blanca, Costa del Sol and the Balearic Islands are visited by 30 million people each year. The tourist industry is now very important to many once poor areas around the Mediterranean, such as Spain, Italy and Greece. Winter holidays are also popular in resorts such as Magaluf. Although the winters are not dry, they are milder than in northern Europe.

 Use photo A and the brochure text to help you write a postcard home. Mention where the resort is; the weather; what there is to see and do; what you think of the place.

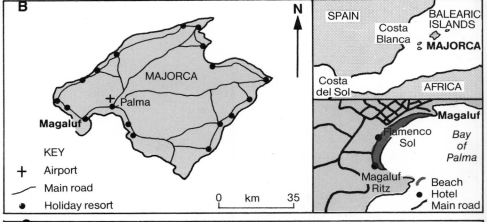

Average daily hours of sunshine

	J	F	M	A	M	J	J	A	S	O	N	D
Magaluf	5.2	6.1	6.8	7.5	9.5	10.2	11.4	10.7	8.0	6.2	5.1	5.0
Manchester	1.8	2.5	3.7	4.3	5.4	6.1	4.9	4.8	3.8	2.8	2.3	1.5

Table 1 Magaluf: Holiday costs and flights

Hotel:	Magaluf Ritz		Flamenco Sol		Airport (Flight times in brackets)	Departure (Day and time)	Return (Day and time)	Add * £
Basic holiday prices in £ per person at different times of year					**Flights to Majorca**			
No. of nights:	7	14	7	14	Gatwick (130 mins)	Tue 1430	Tue 2030	8
						Sat 1500	Sat 2030	–
Time of year						Tue 2300	Wed 0515	–
23 Mar–28 Mar	197	271	184	244		Sat 2300	Sat 2200	–
29 Mar–8 Apr	196	270	182	243				
9 Apr–30 Apr	195	269	181	243	Birmingham (140 mins)		Tue 1635	23
1 May–9 May	191	299	167	260		Tue 1520	Sat 1505	9
10 May–16 May	208	323	190	280		Sat 1520		
17 May–23 May	216	324	195	282				
24 May–30 May	219	326	201	288	Manchester (160 mins)	Sat 1715	Sat 2310	12
31 May–20 Jun	208	314	206	284		Tue 1745	Tue 2330	21
21 Jun–4 Jul	232	372	210	330		Sat 1745	Sat 2035	12
5 Jul–11 Jul	249	398	220	344				
12 Jul–19 Jul	265	412	241	362	Newcastle (170 mins)	Sat 0800	Sat 1430	20
20 Jul–11 Aug	274	418	249	368		Sat 2215	Sat 2130	41
12 Aug–26 Aug	269	413	241	362				
27 Aug–8 Sep	260	404	237	357	Glasgow (175 mins)	Sat 1650	Sat 1550	63
9 Sep–22 Sep	257	400	236	352		Tue 1650	Wed 2015	47
23 Sep–6 Oct	236	339	213	301				
7 Oct–15 Oct	218	325	198	290	Edinburgh (175 mins)	Sat 1440	Sat 1335	23
16 Oct–29 Oct	208	301	193	272		Sat 0830	Sat 1500	59

* Please add airport supplement (right-hand column) to holiday cost.

D Magaluf: holiday paradise

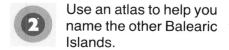

2 Use an atlas to help you name the other Balearic Islands.

3 Study the charts in diagram C then answer the following questions.
(a) Which month is the hottest in Magaluf?
(b) What is the highest temperature?
(c) Which month is driest?
(d) Describe summers in Magaluf.
(e) Describe winters in Magaluf.
(f) Which months are very sunny?

4 Your family has decided to visit Magaluf either in the Easter break (7 days) or in the summer holidays (14 days). Use Table 1 to help you work out
(a) how long it would take to fly to Magaluf from the airport listed which is nearest to where you live;
(b) the cost of an Easter holiday in Magaluf Ritz;
(c) the cost of a summer holiday (leaving 8 July) in Flamenco Sol.

5 (a) What other costs would you need to save for?
(b) Why do the prices vary so much from March to October?

6 What are the advantages and disadvantages of holidaying in a hotel rather than an apartment or a caravan?

7 Make a key for sketch D by writing out the labels below, and matching them to the correct numbers on the sketch.

Beach-front hotels	Water sports

Tideless Mediterranean sea	

Popular crowded beaches	Marina

Old settlement with discos, shops	

Long sandy beach	Clear blue skies

Trees give shade	

43 BREATHING SPACE

A magnet for Alpine fans from all over the world, Zermatt (A) is set high in the beautiful Valais region of southern Switzerland (B). The magnificent Matterhorn towers above the charming village, offering a constant challenge to climbers. The superb skiing, fantastic views and bustling village life make Zermatt a place to visit in all seasons, time and time again.

A Zermatt and the Matterhorn in winter

You can find lots of brochures for mountain holiday resorts. Mountain holidays are the next most popular type after the 'sun, sea and sand' holiday. Europe's mountains are popular for summer and winter holidays, especially the Alps.

Resorts such as Zermatt have many advantages for tourism.
- In winter, the snow-covered mountains provide many ski slopes.
- In summer, the same mountains offer spectacular scenery, ideal for climbing, walking and sight-seeing (photo C).

- The clean air is healthy and there is sunshine in winter as well as summer.
- Hotels are very modern and contain the latest luxurious suites and rooms (such as the Hotel Hofalpen: see the table in B).
- Purpose-built sporting facilities include swimming pools, saunas, squash courts, tennis courts and a solarium.
- There are theatres, concert halls and cinemas.

B Zermatt: winter holiday information

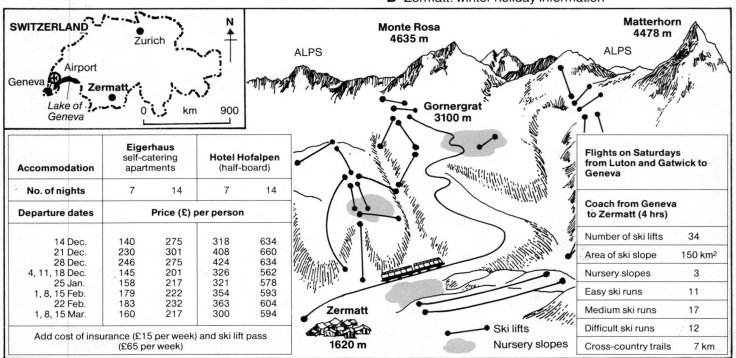

Accommodation	Eigerhaus self-catering apartments		Hotel Hofalpen (half-board)	
No. of nights	7	14	7	14
Departure dates	Price (£) per person			
14 Dec.	140	275	318	634
21 Dec.	230	301	408	660
28 Dec.	246	275	424	634
4, 11, 18 Dec.	145	201	326	562
25 Jan.	158	217	321	578
1, 8, 15 Feb.	179	222	354	593
22 Feb.	183	232	363	604
1, 8, 15 Mar.	160	217	300	594
Add cost of insurance (£15 per week) and ski lift pass (£65 per week)				

Flights on Saturdays from Luton and Gatwick to Geneva

Coach from Geneva to Zermatt (4 hrs)

Number of ski lifts	34
Area of ski slope	150 km²
Nursery slopes	3
Easy ski runs	11
Medium ski runs	17
Difficult ski runs	12
Cross-country trails	7 km

C Summer hill-walking in the Alps

7 Make a key for sketch D by writing the labels below and matching them to the correct numbers on the sketch.

| Mountain holiday resort |

| Cable-cars | | Hill-top restaurant |

| Hang-gliding | | Modern hotels |

| Coach transfer from distant airport |

| Ski lifts and mountain railway |

| Rock climbing |

| Forests on north-facing slopes |

| Lakes: summer – water sport
           winter – skating |

| Alps: hill-walking, cross-country skiing |

1 What advantages do mountain holiday resorts such as Zermatt have for
(a) winter sports people?
(b) young summer visitors?
(c) elderly summer visitors?
Write a sentence for each to explain.

6 **Leisure activities** can be split into active and passive types (see page 88). Make two lists, one of active and one of passive leisure activities available in mountain resorts such as Zermatt.

2 Using the table in B, work out how much it would cost for you and a friend to go on a ski holiday in Zermatt leaving on 21 December for seven nights. You have a choice of accommodation in Zermatt. Decide which is most suitable for you.

3 (a) What are the advantages of staying in a hotel?
(b) What are the advantages of staying in apartments?

4 (a) When are the prices highest?
(b) Why is insurance cover very important on a winter holiday?

5 Which standards of skiers are catered for in Zermatt? (Look at the sketch in B.)

D

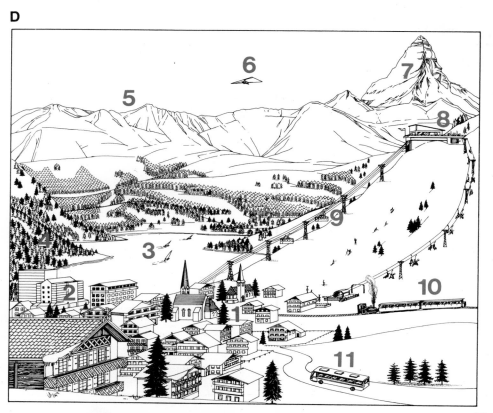

44 Running the country

More and more people are spending their spare time visiting Britain's country parks. In Scotland, there are 34 country parks (see map A). Most of the parks are within easy travelling distance of large cities and are very popular, especially at weekends.

Country parks have become important in the last 20 years, as the demand for more **leisure activities** has increased. More people have more leisure time because of the shorter working week, longer holidays and unemployment. There is also a greater demand for healthy outdoor activities in the fresh air of the countryside.

Popular leisure activities in country parks range from dog walking and picnics to rock climbing and fun-running (see picture B and chart C).

The **amenities** provided in country parks have been improved in recent years. Visitor centres, such as Beecraigs Centre (picture B4) are common. There are now good car parks, toilets, barbecue sites and trim tracks. Some country parks even have specially produced maps (see map D) for orienteering and walking.

The increasing demand for more land for country parks can cause problems. Farmers are against this use of fertile farmland. They complain about the damage to walls and crops from park visitors. Traffic jams are a problem at weekends. Litter from picknickers is also common. Local authorities employ rangers to keep an eye on their country parks. They also provide litter bins and put up notices about using the 'Country Code' (see picture B).

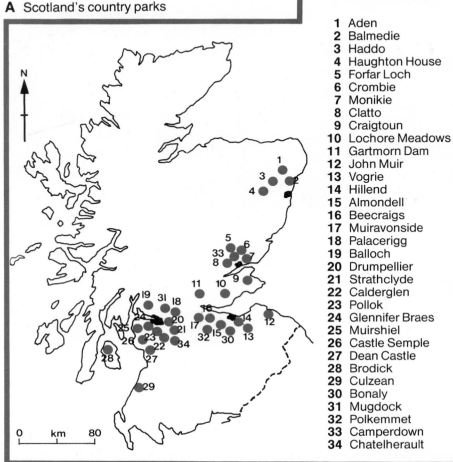

A Scotland's country parks

● **Country parks**

1 Aden
2 Balmedie
3 Haddo
4 Haughton House
5 Forfar Loch
6 Crombie
7 Monikie
8 Clatto
9 Craigtoun
10 Lochore Meadows
11 Gartmorn Dam
12 John Muir
13 Vogrie
14 Hillend
15 Almondell
16 Beecraigs
17 Muiravonside
18 Palacerigg
19 Balloch
20 Drumpellier
21 Strathclyde
22 Calderglen
23 Pollok
24 Glennifer Braes
25 Muirshiel
26 Castle Semple
27 Dean Castle
28 Brodick
29 Culzean
30 Bonaly
31 Mugdock
32 Polkemmet
33 Camperdown
34 Chatelherault

0   km   80

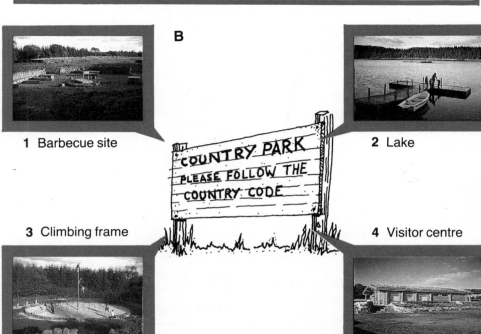

B

1 Barbecue site

2 Lake

3 Climbing frame

4 Visitor centre

COUNTRY PARK PLEASE FOLLOW THE COUNTRY CODE

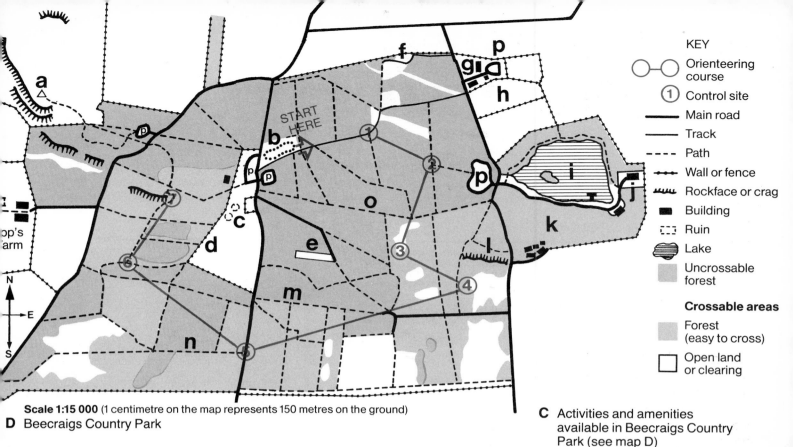

KEY

⚬—⚬	Orienteering course
①	Control site
▬▬	Main road
——	Track
- - -	Path
•–•–•	Wall or fence
⌇⌇⌇	Rockface or crag
▬	Building
⌗	Ruin
≋	Lake
	Uncrossable forest

Crossable areas

| | Forest (easy to cross) |
| | Open land or clearing |

Scale 1:15 000 (1 centimetre on the map represents 150 metres on the ground)

D Beecraigs Country Park

1 Look carefully at map A.
(a) Where are most of Scotland's country parks? Why do you think this is?
(b) Why are there few country parks in north-west and southern Scotland?

2 Why have country parks become more important in recent years?

3 (a) Where is Beecraigs Country Park (number 16 on map A)?
(b) Which type of tree grows there? (Hint: look at the photos in picture B.)
(c) What is unusual about the visitor centre (photo B4)?

4 What problems might there be in Beecraigs Park:
(a) on a dry day with open fires at the barbecue site?
(b) if a scout group takes a short-cut across Kipps farmland (see map D)?
(c) on a sunny weekend afternoon? Who might complain to the park authorities in each case?

5 Study map D. An orienteering course is marked on it.
Complete a larger copy of the table below (for the route through all seven controls) by working out the quickest and easiest route around the course. (Hint: open land, clearings and some parts of the forest (pale red areas) can be crossed quickly. In uncrossable forest you must only travel along paths.)

C Activities and amenities available in Beecraigs Country Park (see map D)

a. Viewpoint (Cockleroy Hill)
b. Trim track/climbing frame
c. Barbecue site/picnics
d. Camping
e. Target archery
f. New campsite
g. Park centre
h. Deer farm
i. Loch: fishing, sailing, canoeing
j. Trout hatchery (fish farm)
k. Field archery
l. Rock climbing
m. Pony trekking trails
n. Nature walks/trails
o. Permanent orienteering course
p. Car park

Controls	Control site	My route would be
Start to control 1	Track/path junction	Run along track to the east
Control 1 to control 2	Path junction	Follow track to the north. Turn east along clearing. Turn right at path and follow it to control 2
⋮		
Control 5 to control 6		
Control 6 to control 7		

45 For better or for worse

The growth of tourism can be a mixed blessing for the places that tourists visit. The cartoons on this page show some of the benefits and some of the disadvantages.

Tourism brings many benefits to holiday areas. Tourists spend money in hotels, shops, discos, restaurants and sport centres. This increases the trade in the tourist centre and it also creates jobs.

As tourist centres develop, so do other things that are linked to the tourist industry. For example, airports, roads and railways may be built or made bigger and better to carry visitors to and from the tourist centres.

A Tourism brings mixed blessings

Pollution is a big problem on popular coasts

Tourism brings employment

Tourism can bring extra noise

Lots of tourists mean more business for shopkeepers

Farmers sell more produce to local hotels

Resorts can be quiet in winter. Many hotels, shops, restaurants close down

Tourism can help to pay for better services such as water, sewage disposal

As holiday resorts grow, they use up surrounding farmland

Overcrowding is a problem in popular resorts

Lots of tourists mean more and better kinds of transport

For many countries, such as Spain and Italy, tourism is a very important part of the economy. It can also be very important to the economy of a single town, such as the seaside resort of Blackpool or the winter sports centre of Aviemore in Scotland.

But tourism can also create problems for tourist areas. For example, streets, airports and beaches can become overcrowded and congested at peak visiting times. Pollution is already a big problem in many Mediterranean resorts. Untreated sewage is pumped into the sea from many hundreds of hotels, holiday flats and houses. Aircraft and cars produce unhealthy fumes. Litter spoils the appearance of beaches and the countryside. Farmland may be bought up and built on as tourist resorts grow.

Visitors often go to holiday areas to see the scenery, relax and 'get away from it all'. But when thousands of people are doing the same thing, the point can be lost!

Country areas are of great importance to the economy. The land may be used for farming, for quarrying, for generating power and water supplies, and for growing forests. Tourism does not mix well with these uses of the land, and a conflict of interest may result. You can see some examples of this conflict in the pictures in B.

B Sign language?

1 Study the text and the cartoons on page 98. Copy the table below, and write at least three examples in each column.

Mixed blessings

Advantages of tourism	Disadvantages of tourism

2 Three kinds of pollution are described in these two pages. Name them. (Look back at page 52 to remind yourself about the different types of pollution.)

3 Look at picture B.
Choose two of the signs, and for each one say
(a) who might have written the sign;
(b) who it is written for;
(c) what might happen if the sign is ignored.

4 Choose three examples of the disadvantages of tourism and suggest ways of solving the problems.

46 For what it's worth

A Survival kit of emergency resources

Imagine you have been shipwrecked on a tropical island, with only the five objects shown in picture A in your 'survival kit'. You find a clearing near a river. It has a small pool in it and is surrounded by trees and bamboo plants. The soil looks good, and there is a large pile of stones near the pool.

The objects in your kit, and the things you find on the island, are the **resources** you can use to help you to survive.

The world, like our tropical island, is full of resources that people can use to help them to survive. Some of these resources, such as wind, air and rain will never run out. Others, such as animals and plants can always be replaced or renewed by nature. These are called **renewable resources**.

There are other resources which cannot be renewed and which will eventually run out. These are called **non-renewable resources** and include such things as coal, iron, oil and tin.

People are using up many of the world's non-renewable resources at a very fast rate. The world's population is growing very quickly, so there is more and more demand for resources. The biggest demand comes from the richer industrial countries such as the U.S.A., France, Britain and Japan. These countries use up vast amounts of the world's resources to make products ranging from ships and bombs to toothbrushes and toys. They also use huge quantities of oil, gas and coal to produce the power needed for industries and to provide fuel for cars, trains, aeroplanes and other forms of transport.

If the world's **consumption** (using up) of resources goes on at the present rate, then some will run out in your lifetime. The coloured bars on chart B show how long eight of the world's resources will last if we do not slow down, or reduce, our consumption. If we can learn to consume the resources more slowly, or if we can conserve them, they will last longer. The complete bars on chart B show how long each resource could last if we try to conserve it.

B How long will they last?

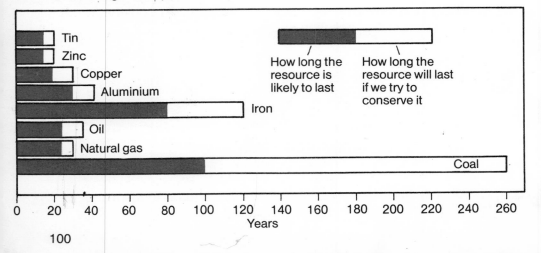

1. If you were stranded on the tropical island, what might you use each of the resources in the 'survival kit' for?

2. Some of the resources you would find on the island are named in the text. What would you use them for?

3. Name a resource which is
   (a) scarce
   (b) plentiful
   (c) cheap
   (d) expensive
   (e) easy to find
   (f) re-usable

4. (a) Are the resources in bar chart B renewable or non-renewable?
   (b) Write down two uses each for four of the resources in chart B.

5. Look at chart B. Which of the resources are likely to run out
   (a) within 20 years' time?
   (b) within 30 years' time?

6. If we use oil faster and faster each year in what year will it run out?

7. Which of the resources in chart B will probably have run out by the time you are 60 years old?

8. How can oil, gas and coal be made to last longer?

100

47 Can't see the wood...

Several thousand years ago, much of the earth's surface was covered in trees. Over the centuries, people have cut down most of these **natural forests** to clear the land for farming, and to provide wood (**timber**) for fuel, buildings and many other things. Only a few areas of natural forest remain in the world today (map A) and some of these are being rapidly used and destroyed.

Nowadays, most of the timber used in industry is specially grown in **commercial forests** (see illustration B). Commercial forestry is very like farming in that there are **inputs** (soil, rain, seed, labour) and **outputs** (timber). The main difference is that trees take much longer to grow than farm crops. It can be over 40 years before a tree is ready to be harvested.

Commercial forests are mostly grown in highland areas where the land is not very suitable for farming.

B Commercial forestry: sitka spruce

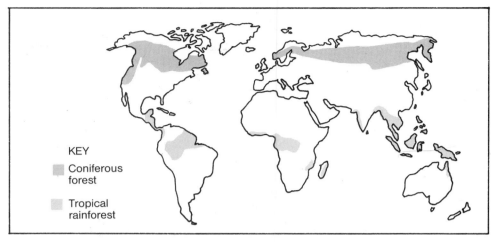

A Main areas of natural forest in the world

KEY
◼ Coniferous forest
▨ Tropical rainforest

C The forestry cycle:
a sitka spruce plantation

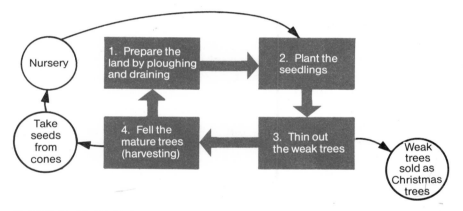

Nursery → 1. Prepare the land by ploughing and draining → 2. Plant the seedlings → 3. Thin out the weak trees → Weak trees sold as Christmas trees → 4. Fell the mature trees (harvesting) → Take seeds from cones → Nursery

Commercial forestry: some facts

Large machines and expensive equipment are needed, for planting and felling trees. But, because of the long time it takes for forests to grow, these can lie unused for long periods.

Forestry does not provide many jobs for people.

Because they take so long to grow, and cover such large areas, forests have a big effect on the environment.

To make a profit, forests must cover a large area. One of the biggest in Britain is Keilder Forest, in north England. It covers 45 000 hectares.

The crop takes a long time to grow. **Softwoods** (such as pine, larch and spruce) take between 40 and 70 years. **Hardwoods** (such as oak, ash and beech) take well over 100 years.

The crop is very bulky. Trees 50 metres high and weighing many tonnes are common in Britain.

To make the most of a valuable **renewable resource** like timber, good management is essential. The seeds must be carefully selected and young seedlings properly protected before being planted out. The ground chosen for planting has to be cleared, ploughed and sometimes drained and treated before the young trees can be planted. Over the next 10–60 years, the forest may be thinned out, sprayed with fertilisers and other chemicals, and protected from animals. Deer eat young trees, and people often start fires which can destroy vast areas of forest in a few days.

After being felled, the trees are stripped of branches and taken to the nearest saw or pulp mill.

The countries of northern Scandinavia produce a large amount of timber (see diagram D), much of which is exported to other countries. The trees that grow in these forests are coniferous trees, such as Scots Pine and Norway Spruce. For many years, the coniferous forests have been commercially forested and earn a great deal of money for Norway, Sweden and Finland. Most of the timber is processed near the forests, either in saw mills or pulp mills, before being exported. Map E shows the position of the main pulp and paper mills in the three countries. Chart F shows the main uses of Scandinavian timber.

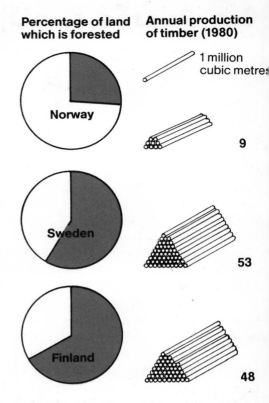

D Timber production in Scandinavia, 1980

E Pulp, paper and saw mills in Scandinavia

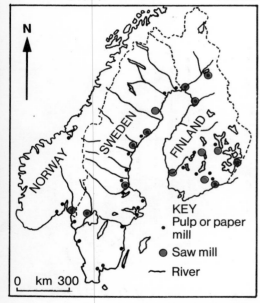

There are a number of threats facing the forestry industry in Europe, and other parts of the world as well.

- Increased felling of timber, but no increase in planting, will be a problem for the future.
- **Acid rain**. Increased air pollution, in the developed world, is slowly killing our forests. Scandinavian forests are affected by pollution from Britain, West Germany and Poland.
- Fires. A severe forest fire can destroy a forest that may have taken 60 years to grow. A constant lookout must be kept, especially in dry countries.

F Timber products in Scandinavia

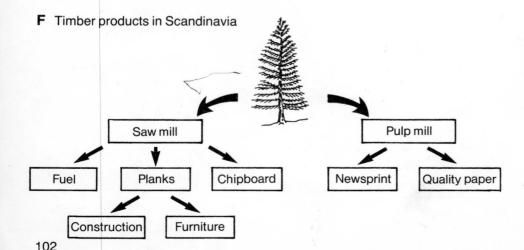

1 Copy and complete the following sentence: 'Timber is a renewable resource because . . .'

2 Use map A and your atlas to name two countries where coniferous forests are found, and two countries where tropical rainforests are found.

3 Explain what the forestry cycle is (diagram C).

4 Use your atlas to help you find out why Norway is less forested than Sweden or Finland.

5 Explain why most of the saw mills and pulp mills in Scandinavia are found by rivers or lakes.

48 Spare a copper

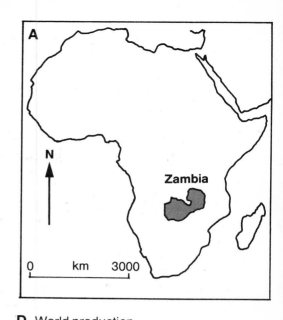

A

Zambia

0   km   3000

Minerals such as coal, oil and copper are all examples of **non-renewable resources**. This means they will eventually become used up (see page 100).

Zambia, in the south of Africa (map A), is the world's fifth largest producer of copper (bar chart D). The copper is found as an **ore** and, after mining, it is processed to produce the pure metal. The mined ore contains 2–6% metal; the rest is waste.

What is going to happen when the copper runs out? One way of slowing down the use of copper is to find alternative materials that do the same job. Already, plastic is being used to make water pipes and telephone fibres, both of which used to be made of copper.

Copper waste can be melted down and re-used. Already over a third of all the copper made each year comes from this waste.

If Zambia's copper runs out, or if people stop buying it, then Zambia will be a much poorer country than it is today. At present, copper exports make up nearly 90% of Zambia's earnings each year (pie chart C).

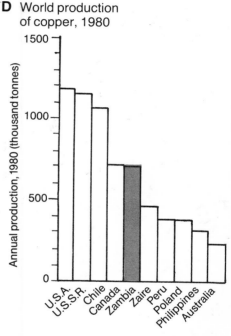

D   World production of copper, 1980

Annual production, 1980 (thousand tonnes)

U.S.A., U.S.S.R., Chile, Canada, Zambia, Zaire, Peru, Poland, Philippines, Australia

B   Ready to blast! A mine in Zambia

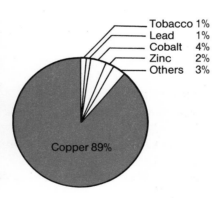

Tobacco 1%
Lead 1%
Cobalt 4%
Zinc 2%
Others 3%

Copper 89%

C   Zambia's main exports by value, 1982

E   The uses of copper

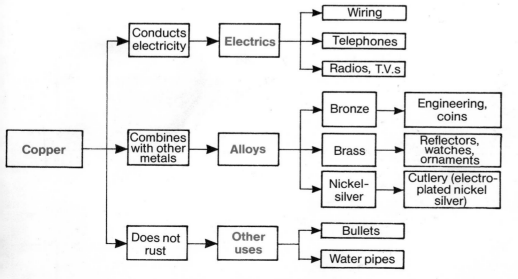

Copper

Conducts electricity → Electrics → Wiring / Telephones / Radios, T.V.s

Combines with other metals → Alloys → Bronze → Engineering, coins / Brass → Reflectors, watches, ornaments / Nickel-silver → Cutlery (electro-plated nickel silver)

Does not rust → Other uses → Bullets / Water pipes

1 Copy and complete the following sentence: 'Copper is a non-renewable resource because . . .'

2 Use chart E to write a paragraph about the importance of copper in the modern world.

3 The price paid for copper changes from day to day. Sometimes the changes can be very great. What problems must this cause for Zambia?

49 THE POWERS THAT BE

If we want to boil a kettle, heat our houses or start off a machine, we need power. For most of us, that power will be electricity.

Electricity is generated using one of two main methods.

1. Water is boiled to produce steam which is used at pressure to turn turbines. The turbines drive generators which make electricity.
   To boil the water, a source of heat is needed such as the sun (photo A), burning coal (photo B) or a nuclear reactor (photo C).
2. The turbines and generators are driven with fast-moving water (photo D) or fast-moving air.

Diagram E shows eight different types of power station or power generator. Each needs energy (heat, water or wind) to drive the turbines. This energy comes from two main types of resource:

(a) **renewable resources**, such as the wind, water and the sun which will never run out;

(b) **non-renewable resources** such as coal, oil and radioactive rocks (e.g. uranium) which will run out one day.

Coal, oil and uranium are types of **fuel**. They have to be burnt or changed in some way to produce heat.

It is not possible to store large amounts of electricity, so it must be manufactured to meet demand. There are over 100 power stations in Britain, all of which are connected to homes and factories throughout the country, as well as to each other. This network of powerlines is called the national grid.

A Solar power station

B Coal-fired power station

C Torness nuclear power station

D Hydro-electric power station

E

Transmission lines normally follow as hidden a route as possible, to avoid being too noticeable (e.g. through valleys, not across a skyline)

In mountainous areas, a dam can be built in a valley to provide a supply of water to produce hydro-electricity

Solar power stations are usually found on slopes where they can collect as much sunshine as possible

Nuclear power stations must be built in fairly remote areas, far from large towns. They are built by the sea or lakes because they need lots of water for cooling

When electricity is cheap (at night) some hydro-electric power stations pump water back up to the high-level reservoir to be re-used next day

The water rushes down pipes to the power station where it turns turbines

In parts of the world where there are large tides the movement of the waves can be used to turn turbines

Oil-fired power stations are usually found beside deepwater to allow oil tankers to dock

Windmills can be built on windy hilltops to produce electricity

Coal-fired power stations are found on flat ground, close to a supply of coal and a supply of water for cooling purposes

Coal stocks can cover large areas of land and are unsightly

When transmission cables would be very noticeable, or would have to cross busy areas, they may be buried below ground, although this is very expensive

Coal mines

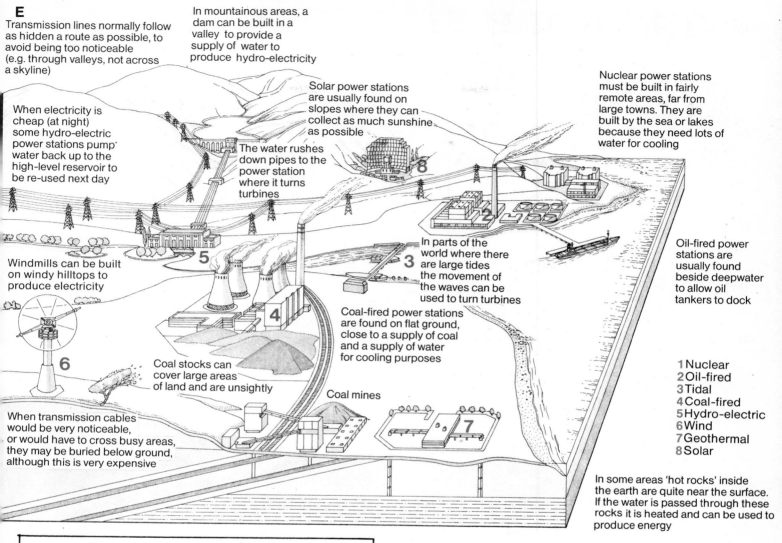

1 Nuclear
2 Oil-fired
3 Tidal
4 Coal-fired
5 Hydro-electric
6 Wind
7 Geothermal
8 Solar

In some areas 'hot rocks' inside the earth are quite near the surface. If the water is passed through these rocks it is heated and can be used to produce energy

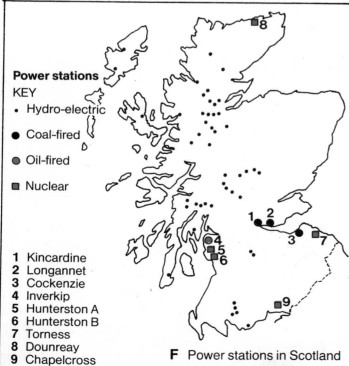

Power stations
KEY
• Hydro-electric
● Coal-fired
◉ Oil-fired
■ Nuclear

1 Kincardine
2 Longannet
3 Cockenzie
4 Inverkip
5 Hunterston A
6 Hunterston B
7 Torness
8 Dounreay
9 Chapelcross

F Power stations in Scotland

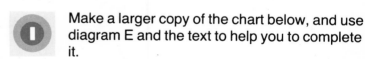

1 Make a larger copy of the chart below, and use diagram E and the text to help you to complete it.

Number on diagram E	Type of power station	Type of energy source: renewable or non-renewable
1	nuclear	non-renewable
2		
3		
⋮		
8		

2 Look at map F and an atlas map of Scotland. Try to explain why the four types of power station are located where they are. Use diagram E to help you.

105

50 Some strike it lucky

One part of the world that is always in the news is the Middle East, often because of its wars and conflicts. We do not hear so much about oil, and yet some Middle Eastern countries produce huge amounts of it (photo A).

Map B shows the main oil-producing countries of the Middle East, and diagram C shows how much oil they produced in 1980. As you can see, the United Kingdom produced as much oil as many Middle Eastern countries in 1980.

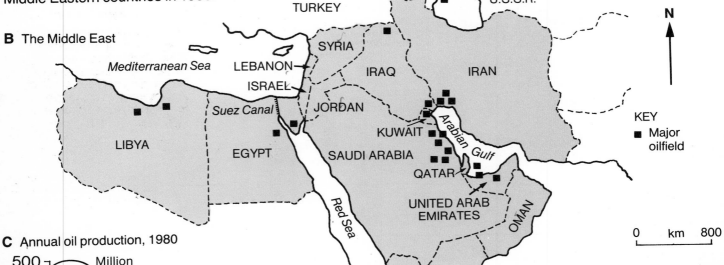

A Middle East oilfield, Oman

B The Middle East

KEY
■ Major oilfield

0   km   800

C Annual oil production, 1980

Million tonnes of oil

Some facts
- Saudi Arabia is the world's second-largest producer of oil (after the U.S.S.R.)
- Until 1980, Iran was the fourth-largest producer, but since a revolution and a war with Iraq, Iran's oil production has been uncertain and exporting has become irregular.

For comparison

Saudi Arabia   Iraq   Libya   United Arab Emirates   Kuwait   Egypt   Qatar   United Kingdom

Kuwait and the other oil-producing countries in the Middle East were all poor countries until they discovered oil about 50 years ago. Now, some of them are very wealthy indeed. One reason for this is because they export most of their oil to other countries, and so earn a great deal of money from the sales.

Some of the oil-producing countries have a very small population for their size. Saudi Arabia, for example, has nine times the area of the United Kingdom, but has only one-seventh of the population (8 million people). This means that Saudi Arabia needs less oil than the United Kingdom, and so can export most of it.

Oil is a **non-renewable resource**. This means it will eventually run out. Experts predict that the Middle East oil supplies could last until the year 2035, but what happens after that?

The huge profits being made from oil make it possible for countries like Oman and Kuwait to plan for their future.

As this page shows, the money can be spent in many different ways such as building new industries, improving farming, and building hospitals, roads and houses. In this way, when the oil runs out, there will be other sources of wealth.

IRAQ
- New motorway being built
- New power station being built
- A great deal of new housing being built

EGYPT has recently begun to develop oilfields in the Red Sea. Some of the profits are being spent on improving farming and medical care.

The wealthiest Arab countries (Saudi, Kuwait, Libya, Iraq) donate foreign aid to the poorer Arab countries such as the Yemens, Jordan and Syria, which have little or no oil.

How the money is spent

Many Arab countries have spent massive amounts on defence. Saudi Arabia, Iraq, Iran and Libya in particular have very large armed services although in the case of Iraq and Iran they spend most of their time fighting each other.

D

Map labels: U.S.S.R., TURKEY, Mediterranean Sea, SYRIA, LEBANON, ISRAEL, IRAQ, IRAN, Suez Canal, JORDAN, LIBYA, EGYPT, KUWAIT, Arabian Gulf, SAUDI ARABIA, QATAR, UNITED ARAB EMIRATES, OMAN, Red Sea, YEMEN ARAB REPUBLIC, PEOPLE'S DEMOCRATIC REPUBLIC OF YEMEN

SAUDI ARABIA

With more than a million pounds being earned every hour by oil, there is no shortage of money to spend on improvements! At present, the government is concentrating on three things:

1 Social welfare
more hospitals, more houses and more university places. Saudi spends more on education than any other country in the world.

2 Industrial development
new factories are being built, especially those which process oil. Foreign experts are paid to train local workers.

3 New schemes
for agriculture, especially irrigation, and soil improvements; also water pipelines and railway line.

1 Write a paragraph to explain why some Middle East countries 'struck it lucky'.

2 Look at map B. List (a) the Middle Eastern countries which are major oil producers, and (b) those which are not.

3 Read the text in the boxes on this page, and then describe the main ways in which Middle East oil countries are 'planning for the future'.

4 The Suez Canal (marked on map B) is used to transport oil from many Middle Eastern countries to Europe. When it was closed during a war between Israel and Egypt in 1956, this was a great blow to the oil companies. Why? (Hint: Study an atlas map of the Middle East and Europe.)

5 Look at diagram C. Which country has the most control over oil prices. Why do you think this is?

51 Over and over

Water is one of the world's most valuable **resources**. People need it to drink, to grow plants, to make power and to transport things. The use of water supplies must therefore be carefully planned to make sure all these needs can be met. Large rivers provide a big supply of water which can be used in many different ways, so they are particularly useful.

The River Rhône (map A) has its source in the French Alps. There is a lot of rainfall in these mountains, and snow in winter which melts later in the year. This makes the Rhône a large river, even when it flows through the south of France which is hot and dry. Water from the Rhône is used:

- to irrigate farmland
- to make hydro-electric power
- to improve navigation
- for drinking water
- for fishing
- for recreation
- for industry

To allow all these different uses, several water schemes have been set up on the river. These enable the water to be used over and over again because the water is returned to the river after use.

The Donzère–Mondragon Scheme, opened in 1952, is one of the largest in the world (see sketch B and map C). Eventually, more than 300 000 hectares of land along the River Rhône will be supplied with irrigation water for crops. This means many more crops can be grown.

A

B An artist's view of the Donzère–Mondragon Scheme

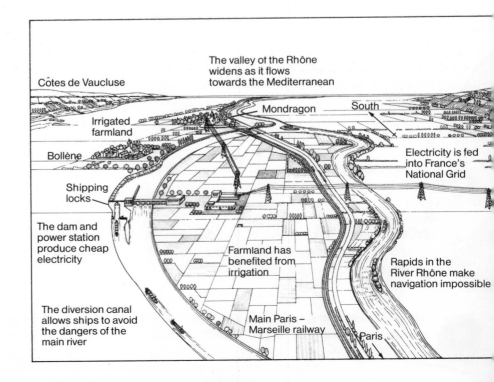

Côtes de Vaucluse

The valley of the Rhône widens as it flows towards the Mediterranean

Mondragon

South

Irrigated farmland

Bollène

Electricity is fed into France's National Grid

Shipping locks

The dam and power station produce cheap electricity

Farmland has benefited from irrigation

Rapids in the River Rhône make navigation impossible

The diversion canal allows ships to avoid the dangers of the main river

Main Paris – Marseille railway

Paris

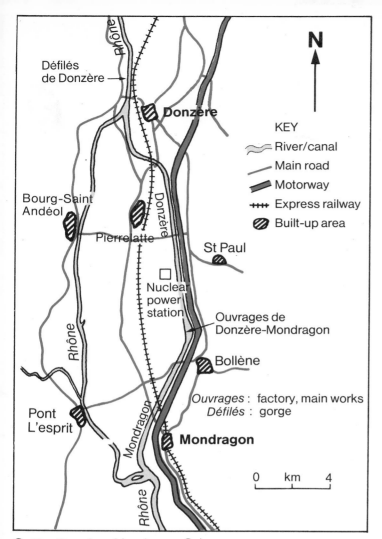

C The Donzère–Mondragon Scheme

KEY
∿ River/canal
— Main road
▰ Motorway
+++ Express railway
▨ Built-up area

Ouvrages : factory, main works
Défilés : gorge

0   km   4

D A river barge on the Rhône

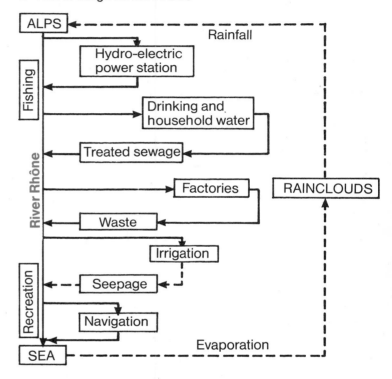

E The River Rhône and the water cycle

In time, twenty power stations will produce electricity for the growing industrial areas of southern France. Ten special navigation channels will also have been built to allow shipping to avoid dangerous stretches of the river.

Flow chart E shows the complete water system of the River Rhône. At each loop in the chart, water is removed for use and then returned to the river. When it reaches the sea, the water evaporates into the air and eventually falls as rain again. In this way it forms a **cycle.**

Trying to manage a resource such as water is very difficult. Human, farming and industrial uses can all cause water pollution (see chart E).

1 Use flow chart E to write a description of the way in which the water of the River Rhône is used.

2 Is water in a river a renewable or non-renewable resource? Explain your answer.

3 Look at sketch B and map C. Why was a diversion canal built on the Donzère–Mondragon Scheme?

4 Different people may disagree about how to use the water in the River Rhône. What disagreements might arise between:
(a) shippers and power station engineers;
(b) farmers and factory owners?

5 'A glass of water drunk in London has, on average, been drunk five times already.' What do you think this means?

52 FAIR EXCHANGE

Look at the shopping list on this page (A). You can buy all these items in British shops but some of them had to be bought by Britain from other countries. The goods that one country buys from another are called **imports**. To help to pay for imports, most countries sell or **export** goods to other countries. This buying and selling of goods is called **trading**.

When the money a country makes from its exports is more than the amount it spends to buy imports, we say that it has a **trade surplus**. If it spends more on imports than it earns from exports it has a **trade deficit** (look at the diagram in B).

A good example of a country which has changed its **balance of trade** is Zambia, in Africa. Look at the pie charts in B. They show Zambia's main exports and imports. The red line on the graph shows how much money Zambia made from its exports each year from 1965 to 1983. The black line shows how much money was spent on imports. As you see, until 1982 the cost of imports was less than the value of exports. This meant that Zambia had a trade surplus until 1982. Then it had a trade deficit.

A

apples
bananas
potatoes
coffee
tea-bags
Danish Blue cheese
currants
marmalade
toilet paper
olive oil

B Zambia: trade figures

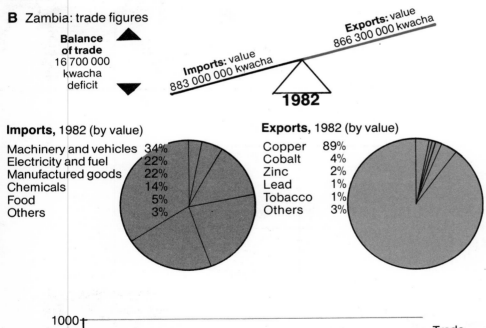

Balance of trade
16 700 000 kwacha deficit

Imports: value 883 000 000 kwacha

Exports: value 866 300 000 kwacha

1982

Imports, 1982 (by value)

Machinery and vehicles	34%
Electricity and fuel	22%
Manufactured goods	22%
Chemicals	14%
Food	5%
Others	3%

Exports, 1982 (by value)

Copper	89%
Cobalt	4%
Zinc	2%
Lead	1%
Tobacco	1%
Others	3%

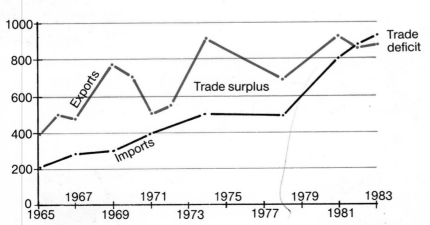

Look again at Zambia's exports and imports. As you can see, Zambia's main exports are **raw materials**. Its imports are mostly **manufactured goods**. A country like Britain shows almost the opposite picture. Britain imports many raw materials and its exports are mostly manufactured goods.

It is because of these differences between the needs of countries that many of them have come together to form special trading groups. In Europe, thirteen countries belong to the European Economic Community (E.E.C.). One of the main aims of the E.E.C. is to allow its members to trade easily and cheaply with one another.

Map D shows another trading group called COMECON. The seven members of COMECON are all communist countries and neighbours of the U.S.S.R.

① Make a chart like the one below. In the first column write the names of all the items listed in shopping list A. Now complete the other two columns for each item. One has been done for you.

Shopping list item	Imported into or produced in Britain	Possible country of origin
Coffee	Imported	Brazil

② Look at map D. With the help of an atlas, list the names of the countries that belong to
(a) the E.E.C.,
(b) COMECON.

③ Look at E. The car in the picture was made in a British factory. But can we call it a British car? Write what *you* think.

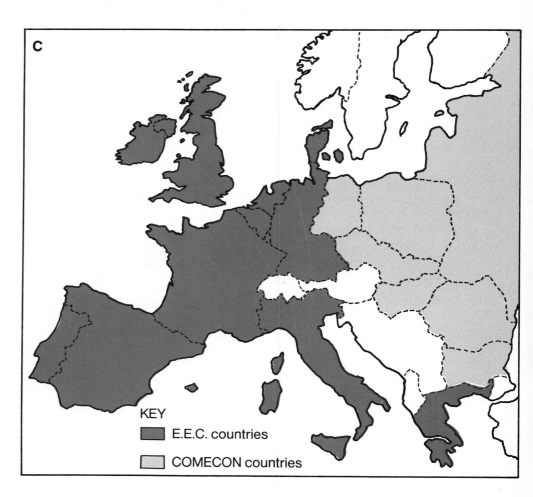

C

KEY
- ▓ E.E.C. countries
- ░ COMECON countries

D The Vauxhall Astra: towards a world car?

Vauxhall and Opel are part of General Motors of Detroit, U.S.A.

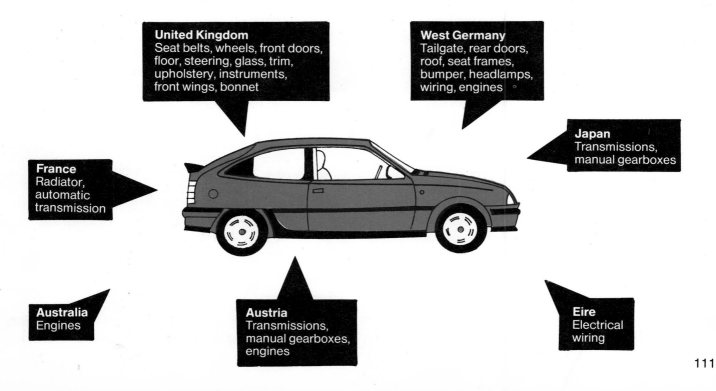

United Kingdom
Seat belts, wheels, front doors, floor, steering, glass, trim, upholstery, instruments, front wings, bonnet

West Germany
Tailgate, rear doors, roof, seat frames, bumper, headlamps, wiring, engines

Japan
Transmissions, manual gearboxes

France
Radiator, automatic transmission

Australia
Engines

Austria
Transmissions, manual gearboxes, engines

Eire
Electrical wiring

53 For richer... for poorer

A

Countries that Unilever does business in

A **multinational** is a company which makes or sells a product in more than one country. As the trade between the countries of the world has grown so have many very large industrial companies. There are now well over 7000 multinational companies, some of which are very large indeed.

Some of the largest multinationals earn more money each year than many poorer countries do. Exxon, the world's largest company, earns more than 100 000 million dollars each year. Only four of the poorer countries in the world earn more than that. The very large multinationals, such as Exxon, General Motors, I.B.M. and Unilever, may be more powerful than many of the countries they have factories in.

The biggest 300 multinational companies are based in the world's richer countries: 200 are American, 80 are European, and 20 are Japanese. The poorer countries need to trade with these large multinationals, but because they are poor they do not always get a good deal. Even the richer countries can lose out. British Petroleum (B.P.), one of the richest multinationals, paid no tax in Britain in 1970.

Many multinationals grow by taking over other smaller companies which make different products. Unilever, the multinational shown in map A, is involved in making many products including margarine, tea, coffee, soaps and detergents, shampoos and toothpastes, soups, tinned and other foods.

E Unilever products

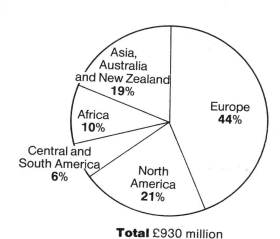

Asia,
Australia
and New Zealand
19%

Africa
10%

Central and
South America
6%

Europe
44%

North
America
21%

Total £930 million

B Where Unilever's profits came from, 1984

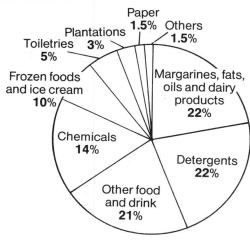

Paper
1.5%   Others
1.5%

Plantations
3%

Toiletries
5%

Frozen foods
and ice cream
10%

Chemicals
14%

Margarines, fats,
oils and dairy
products
22%

Detergents
22%

Other food
and drink
21%

Total £930 million

C What Unilever's profits came from, 1984

D The world's top fifteen companies

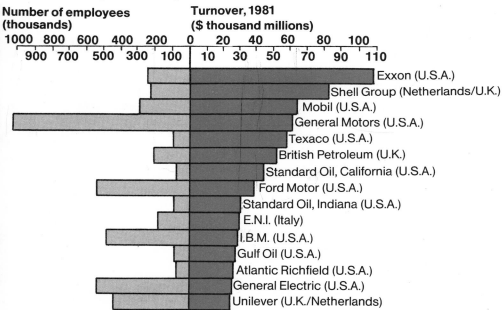

**Number of employees
(thousands)**

1000 800 600 400 200 0
  900 700 500 300 100

**Turnover, 1981
($ thousand millions)**

0   20   40   60   80   100
  10   30   50   70   90   110

Exxon (U.S.A.)
Shell Group (Netherlands/U.K.)
Mobil (U.S.A.)
General Motors (U.S.A.)
Texaco (U.S.A.)
British Petroleum (U.K.)
Standard Oil, California (U.S.A.)
Ford Motor (U.S.A.)
Standard Oil, Indiana (U.S.A.)
E.N.I. (Italy)
I.B.M. (U.S.A.)
Gulf Oil (U.S.A.)
Atlantic Richfield (U.S.A.)
General Electric (U.S.A.)
Unilever (U.K./Netherlands)

1 What is a multinational
company?

2 (a) Look at map A. How
many countries does
Unilever do business in?
With the help of an atlas, make a list
of the countries.
(b) Describe the spread of countries
that Unilever does business in.

3 Why do you think it is
difficult for poorer countries
to get a good deal with a
large multinational?

4 Why do you think
multinationals do business
in more than one country?

54 Moordale's for me!

Turn to pages 58 and 59 and look at the sketch of Moordale. This is an imaginary landscape, but it is very like lots of real places in Britain.

Now look at the five newspaper cuttings on this page (A–E). Each cutting describes a plan to develop part of the Moordale area. It also gives some of the arguments for and against the plan. All the plans will have a big effect on the area, and all of them will please some people but upset others.

What you and your class have to do is to decide which plan would be best for Moordale. Remember that no plan will suit everyone, but it is up to you to argue a good case for going ahead with one of them.

Now read the list of things to do (1–7).

A Hi-Tech boost for Moordale!

Today the government hinted that they would welcome plans for a new industrial estate in Moordale, despite opposition from local farmers. When asked what exactly would be involved, a government spokesperson said, "A new industrial estate covering 450 hectares and linked to local motorways would provide over 600 jobs in hi-tech industries when complete."

But farmers are very worried about the threat to farmland. Tom Brown, a local farmer, is angry. "More valuable farmland will be smothered by industry," he said.

Local conservationists are against any new plans as they would lead to draining marshland and disturbing local wildlife.

B Skiers in Moordale?
£25 million winter sports centre planned

Joan Campbell, chairwoman of the Skears Group Ltd, outlined her company's plans for a new winter sports centre. "There will be 750 new jobs in the Moordale Area," she said. "There will be three major ski slopes within 4 kilometres of the new Skears centre." The company plan to build four hotels, an ice rink, swimming pools, shops, discos and a cinema in the sports centre.

But some locals are not happy. "Opening up Moordale to 5 000 skiers a day will be a disaster!" says Stuart Duncan, a local conservationist. "Apart from traffic congestion in winter, plants will be destroyed and the hillside will be scarred and eroded!" Catherine Ross welcomed the news of jobs for Moordale on behalf of the local job centre.

 Copy and complete this table by reading the newspaper stories A–E.

New development plan	Number of long-term jobs	Groups in favour of new plans	Groups against new plans
A Industrial estate			
B			
C			
D			
E			

2 Form into five groups in your class. Each group will be asked to represent one of the development plans for Moordale. Read the newspaper cutting which describes your plan, and look again at pages 58–9. Discuss your plan in your group.

3 Your group must decide which is the best location for your development. Choose one of the 14 locations marked 1–14 on the Moordale landscape (pages 58–9).

C

Flood of Tears

Part of Moordale is to be flooded if the United Electricity Board gets the go-ahead from local council planners. Proposals to develop the Moordale hydro-electric scheme include the building of a dam and a large reservoir. A generating station linked by pylons to the National Grid will also be built. Construction will provide 600 jobs and will take over four years to complete.

Alan Craig, local Forestry Officer, is disgusted. "If the scheme goes ahead, many local forests and their wildlife will be lost and with them some 65 to 70 local jobs," he said yesterday. Moordale's farmers also stand to lose fertile land and there is some threat to local coal mines.

But the council's planning department see possible benefits for Moordale. "In the end we may attract tourists such as water sports enthusiasts to use the reservoir," said Judy Brack, a planning department official.

D

The Last Resort?

Residents in a small fishing village in Moordale are today confused and angry. Rumours suggest that *Sherlock Homes*, a large building company, is about to get planning permission for a new holiday complex. Plans are said to include new hotels, restaurants, sports facilities, golf courses and holiday time-share apartments. Many local business people are already looking forward to the increase in trade. Over 200 jobs could be provided in the new holiday village.

But locals fear the increase in noise and congestion. Jim Ace, local lobster fisherman, is worried. "If this plan goes ahead, there will be no room in the harbour for fishing boats – only yachts," he said.

Local estate agents said today that house and land prices would rise dramatically if the *Sherlock Homes* complex is built.

E

Nuclear leak – more power to the people!

Controversial plans to build a nuclear power station on Moordale's popular coast were leaked to *The Herald* last night. The plans consider several possible sites for a nuclear plant, but the most favoured sites are those close to the fresh water and flat land. If Moordale is chosen, 600 new jobs would be created and existing road and rail links would be improved.

Local fishermen told *The Herald* they are shocked by the proposals. They say that Moordale Bay could be closed to fishing because of discharged waste from the power station.

A spokesperson for the local branches of the NUM said that the plans pose a big threat to the jobs of Moordale's 1200 coal miners.

CRAM, the anti-nuclear campaign group, is already planning to hold a massive protest march to the Moordale site.

4 Your group must produce a two-page report on your plan. The report must say something about these three things:

(a) Your choice of location for the development. Say why you choose it, and draw a simple sketch or map of the chosen location.

(b) The advantages of your plan for Moordale. Why should your plan go ahead?

(c) Your response to the protestors who will be against your plan. What would you say to them?

5 Each group must choose one person to speak to the class. That person should read the group report to the class.

6 The whole class must now discuss all the reports. When all the arguments are over, the class must vote for one of the five plans. You must decide which is the best plan for Moordale.

7 After the final decision has been made, each group should make a news report for the 'Radio Moordale' evening news programme. The report should include the main reasons for the final choice of development plan and the reasons for its location. It must also include comments made by local people.

Index